AN INFORMATION APPROACH TO MITOCHONDRIAL DYSFUNCTION

Extending Swerdlow's Hypothesis

World Scientific Series in Information Studies
(ISSN: 1793-7876)

Series Editor: Mark Burgin *(University of California, Los Angeles, USA)*

International Advisory Board:

Søren Brier *(Copenhagen Business School, Copenhagen, Denmark)*
Tony Bryant *(Leeds Metropolitan University, Leeds, United Kingdom)*
Gordana Dodig-Crnkovic *(Mälardalen University, Eskilstuna, Sweden)*
Wolfgang Hofkirchner *(ICT&S Center, University of Salzburg, Salzburg, Austria)*
William R King *(University of Pittsburgh, Pittsburgh, USA)*

World Scientific Series in Information Studies — **Vol. 4**

AN INFORMATION APPROACH TO MITOCHONDRIAL DYSFUNCTION

Extending Swerdlow's Hypothesis

Rodrick Wallace

Columbia University, USA

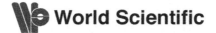 **World Scientific**

NEW JERSEY · LONDON · SINGAPORE · BEIJING · SHANGHAI · HONG KONG · TAIPEI · CHENNAI

Published by

World Scientific Publishing Co. Pte. Ltd.

5 Toh Tuck Link, Singapore 596224

USA office: 27 Warren Street, Suite 401-402, Hackensack, NJ 07601

UK office: 57 Shelton Street, Covent Garden, London WC2H 9HE

Library of Congress Cataloging-in-Publication Data
Wallace, Rodrick, author.
 An information approach to mitochondrial dysfunction : extending Swerdlow's hypothesis / by Rodrick Wallace.
   p. ; cm. -- (World Scientific series in information studies ; volume 4)
 Includes bibliographical references and index.
 ISBN 978-9814663502 (hardcover : alk. paper)
 I. Title. II. Series: World Scientific series in information studies ; v. 4. 1793-7876
 [DNLM: 1. Mitochondrial Diseases--physiopathology. 2. Information Theory. 3. Mitochondrial Degradation. 4. Models, Biological. 5. Statistics as Topic. WD 200.5.M6]
 RB155
 616'.042--dc23
                                                     2015001066

British Library Cataloguing-in-Publication Data
A catalogue record for this book is available from the British Library.

Printed in Singapore

Preface

Speculations and commentary on the relation between information theory and biology range from Quastler's classic 1954 volume to the most current literatures. Almost at random, Brennan et al. (2012) find that information theory allows analyses of cell signaling capabilities without necessarily requiring detailed knowledge of the signaling networks. Adami (2012) examines the use of information theory in evolutionary biology, Schneider (2010) gives an overview of molecular information theory, Wallace (2014c) uses the theory to explore evolutionary 'Cambrian explosions', and so on and on.

Here, we shall combine information theory and control theory via the Data Rate Theorem, and apply information-theoretic 'hidden symmetry' arguments to explore the central role of available metabolic free energy in physiological processes.

There is an important context to this attempt.

First, Swerdlow's mitochondrial cascade hypothesis for Alzheimer's disease (AD) (Swerdlow et al. 2010) proposes that a person's genes determine their baseline mitochondrial function and durability. While both parents influence one's lifetime AD risk, since mtDNA is maternally inherited, mothers have a greater impact than fathers. It is generally accepted that mitochondrial function declines with age, and data suggest this drives a variety of age-associated physiological changes. It is likely cell physiology initially compensates for and adapts to this change, but eventually a point is reached at which adequate compensation is no longer possible. A generalized mitochondrial cascade hypothesis, following Swerdlow's lead, proposes that a person's genetically determined mitochondrial starting line, in conjunction with their genetically and environmentally determined rate of mitochondrial decline, determines the age at which a broad spectrum of clinical diseases ensue.

A second line of argument goes in precisely the other direction: Rossignol and Frye (2010, 2012) collate evidence of mitochondrial dysfunction in autism spectrum disorders, associated with neurodevelopmental pathologies driven by a metabolic energy deficit (Giulivi et al. 2010; Go et al. 2014). Palmieri and Persico (2010) argue that mitochondrial function may play a critical role not just in rarely causing disease, but also in frequently determining to what extent different prenatal triggers will derange neurodevelopment and yield abnormal postnatal behavior, presumably through failure of essential and highly energy-dependent regulatory mechanisms.

Another, similar, mitochondrial dysfunction model can be applied to schizophrenia, which can be seen as a later stage 'developmental' disorder, often associated with postadolescent neural pruning (Ben-Shachar 2002; Parbakaran et al. 2004; Shao et al. 2008; Scaglia 2010; Clay et al, 2011).

More generally, natural cognitive systems operate at all scales and levels of organization of biological process (e.g., Wallace, 2012a, 2014a). The failure of low level biological cognition in humans is often expressed through early onset of the intractable chronic diseases of senescence (e.g., Wallace and Wallace, 2010, 2013). Failure of high order cognition in humans has been the subject of intensive scientific study for over two hundred years, with little if any consensus. As Johnson-Laird et al. (2006) put it,

> Current knowledge about psychological illnesses is comparable to the medical understanding of epidemics in the early 19th Century. Physicians realized that cholera, for example, was a specific disease, which killed about a third of the people whom it infected. What they disagreed about was the cause, the pathology, and the communication of the disease. Similarly, most medical professionals these days realize that psychological illnesses occur... but they disagree about their cause and pathology.

As the media chatter surrounding the release of the latest official US nosology of mental disorders – the so-called 'DSM-V' – indicates, this may be something an understatement. Indeed, the entire enterprise of the *Diagnostic and Statistical Manual of Mental Disorders* has been characterized as 'prescientific' (e.g., Gilbert, 2001). Atmanspacher (2006), for example, argues that formal theory of high-level cognition is itself at a point like that of physics 400 years ago, with the basic entities and the relations between them yet to be determined. Further complications arise via the overwhelming influence of culture on both mental process and its dysfunction (e.g., Heine, 2001; Kleinman and Cohen 1997), something to which we will return. Chapter 5 of Wallace and Wallace (2013) provides a more detailed exploration.

It seems increasingly clear that the stabilization and regulation of high order cognition is as complex as such cognition itself.

Some simplification, we will show, is possible. High level cognition can be described in terms of a sophisticated real-time feedback between interior and exterior, necessarily constrained, as Dretske (1994) has noted, by the asymptotic limit theorems of information theory:

> Communication theory can be interpreted as telling one something important about the conditions that are needed for the transmission of information as ordinarily understood, about what it takes for the transmission of semantic information. This has tempted people... to exploit [information theory] in semantic and cognitive studies...
>
> ...Unless there is a statistically reliable channel of communication between [a source and a receiver]... no signal can carry semantic information... [thus] the channel over which the [semantic] signal arrives [must satisfy] the appropriate statistical constraints of information theory.

Intersection of that theory with the formalisms of real-time feedback systems – control theory – provides insight into matters of embodied cognition and the parallel synergistic problem of embodied regulation and control.

We invoke powerful statistical and mathematical tools in the study of these phenomena, taking seriously a remark by the mathematician John Kemeny to the effect that scientists continually rediscover and reapply well-known mathematical results using childish methods. Here, we begin with those results.

Chapter 1 introduces basic ideas from information and control theory and related matters. Chapter 2 focuses heavily on the 'hidden symmetries' in nonequilibrium phase transitions, first examining basic protein conformations that collapse into amyloid formation, and then the extensions associated with cognitive control and regulation of complex cellular biochemical phenomena. Chapter 3 looks at nonequilibrium phase transitions using a Black-Scholes 'metabolic cost' analysis, generalizing the methods of Tishby and Polani (2011), finding a biochemical analog to the Data Rate Theorem that connects information and control theories, and extending the argument to a broad class of basic biological 'retinas'. Chapter 4 reexamines some of the questions raised in Chapter 3 using mutual information. Chapter 5 studies second-order fragmentation effects that may be important in neurodevelopmental disorders. Chapter 6 extends the model to surprisingly canonical forms of cognitive failure that may express themselves at different levels of organization. Chapter 7 explores the influences of environment and embodiment from these perspectives. Chapter 8 describes how excess demand for metabolic free energy is itself pathological, constituting a generalized inflammation that can lead to premature aging, and looks at the relation between AD and job stress, using data on the disorder from 'right-to-work' (RTW) and non-RTW states in the USA. The final chapter places these results in the context of the failure of molecular biology and other reductionist approaches to cash out on their considerable intellectual and financial investments, as characterized by the 'decline in pharmaceutical industry productivity' indexed by the exponentially increasing cost of bring new drugs to market shown in figure 9.1.

Although it might be possible to take a physics-like 'statistical mechanics' perspective on these matters, we prefer to view the outcome of this analysis as providing a set of statistical models, not unlike dynamic regression equations, that can be fitted to data. Such tools are suitable for comparing similar systems under different, and different systems under similar, experimental or observational conditions. As the mathematical ecologist E.C. Pielou has noted (Pielou 1977, pp. 107-110), the principal use of mathematical models in the biological sciences is for raising hypotheses for data-based study. It is, after all, quite unrealistic to expect that mathematical models of complex biological, ecosystem, or social phenomena will often be 'real' in the sense that analytical mechanics, Maxwell's equations, Einstein's special relativistic and covariant gravitational equations, the Schroedinger and Dirac equations, or the Standard Model, provide detailed, predictive descriptions of what are, in comparison, relatively simple physical processes.

The mathematical level is at the advanced undergraduate or first year graduate student level, with some specialized material provided in an appendix.

The formal development is, from a certain perspective, quite surprisingly direct. By contrast, alternatives to failing 'magic bullet' interventions, as explored in Chapter 8, face deep cultural constraints.

Contents

Chapter 1

Mathematical preliminaries

We begin with some standard results from information theory, make a significant extension based on the homology between information and free energy noted by Feynman (2000) and others that will prove useful in later chapters, and end with a statement of the Data Rate Theorem that connects information and control theories.

1.1   The coding and tuning theorems

Messages from a source, seen as symbols x_j from some alphabet, each having probabilities P_j associated with a random variable X, are 'encoded' into the language of a transmission channel, a random variable Y with symbols y_k, having probabilities P_k, possibly with error. Someone receiving the symbol y_k then retranslates it (without error) into some x_k, which may or may not be the same as the x_j that was sent.

More formally, the message sent along the channel is characterized by a random variable X having the distribution $P(X = x_j) = P_j$, $j = 1, ..., M$.

The channel through which the message is sent is characterized by a second random variable Y having the distribution $P(Y = y_k) = P_k$, $k = 1, ..., L$.

Let the joint probability distribution of X and Y be defined as $P(X = x_j, Y = y_k) = P(x_j, y_k) = P_{j,k}$, and the conditional probability of Y given X as $P(Y = y_k | X = x_j) = P(y_k | x_j)$.

Then the Shannon uncertainty of X and Y independently and the joint uncertainty of X and Y together are defined respectively as

$$H(X) = -\sum_{j=1}^{M} P_j \log(P_j)$$

$$H(Y) = -\sum_{k=1}^{L} P_k \log(P_k)$$

$$H(X,Y) = -\sum_{j=1}^{M}\sum_{k=1}^{L} P_{j,k} \log(P_{j,k}) \tag{1.1}$$

The *conditional uncertainty* of Y given X is defined as

$$H(Y|X) = -\sum_{j=1}^{M}\sum_{k=1}^{L} P_{j,k} \log[P(y_k|x_j)] \tag{1.2}$$

For any two stochastic variates X and Y, $H(Y) \geq H(Y|X)$, as knowledge of X generally gives some knowledge of Y. Equality occurs only in the case of stochastic independence.

Since $P(x_j, y_k) = P(x_j)P(y_k|x_j)$, then $H(X|Y) = H(X,Y) - H(Y)$.

The information transmitted by translating the variable X into the channel transmission variable Y – possibly with error – and then re-translating without error the transmitted Y back into X is defined as $I(X;Y) \equiv H(X) - H(X|Y) = H(X) + H(Y) - H(X,Y)$. See Cover and Thomas (2006) for details. If there is no uncertainty in X given the channel Y, then there is no loss of information through transmission. In general this will not be true, and in that is the essence of the theory.

Given a fixed vocabulary for the transmitted variable X, and a fixed vocabulary and probability distribution for the channel Y, we may vary the probability distribution of X in such a way as to maximize the information sent. The capacity of the channel is defined as $C \equiv \max_{P(X)} I(X;Y)$, subject to the subsidiary condition that $\sum P(X) = 1$.

The critical trick of the Shannon Coding Theorem for sending a message with arbitrarily small error along the channel Y at any rate $R < C$ is to encode it in longer and longer 'typical' sequences of the variable X; that is, those sequences whose distribution of symbols approximates the probability distribution $P(X)$ above which maximizes C.

If $S(n)$ is the number of such 'typical' sequences of length n, then $\log[S(n)] \approx nH(X)$, where $H(X)$ is the uncertainty of the stochastic variable defined above. Some consideration shows that $S(n)$ is much less than the total number of possible messages of length n.

As $n \to \infty$, only a vanishingly small fraction of all possible messages is meaningful in this sense. This, after some development, allows the Coding Theorem to work so well. The prescription is to encode messages in typical sequences, which are sent at very nearly the capacity of the channel. As the encoded messages become longer and longer, their maximum possible rate of transmission without error approaches channel capacity as a limit.

The argument can, however, be turned on its head, in a sense, to provide a 'tuning theorem' variant to the coding theorem.

Telephone lines, optical wave guides and the tenuous plasma through which a planetary probe transmits data to earth may all be viewed in traditional information-theoretic terms as a noisy channel around which it is necessary to structure a message so as to attain an optimal error-free transmission rate. Telephone lines, wave guides and interplanetary plasmas are, however, fixed on the timescale of most messages, as are most sociogeographic networks. Indeed, the capacity of a channel is defined by varying the probability distribution of the message X so as to maximize $I(X;Y)$.

Assume, now, however, that some message X so critical that its probability distribution must remain fixed. The trick is to fix the distribution $P(x)$ but now to modify the channel, to tune it, to maximize $I(X;Y)$. A dual channel capacity C^* can be defined as $C^* \equiv \max_{P(Y),P(Y|X)} I(X;Y)$.

But $C^* = \max_{P(Y),P(Y|X)} I(Y;X)$ since $I(X;Y) = H(X) + H(Y) - H(X,Y) = I(Y;X)$.

Thus, in a formal mathematical sense, the message 'transmits the channel', and there will be, according to the Coding Theorem above, a channel distribution $P(Y)$ which maximizes C^*.

Thus modifying the channel may be a far more efficient means of ensuring transmission of an important message than encoding that message in a 'natural' language which maximizes the rate of transmission of information on a fixed channel.

We have examined the two limits in which either the distributions of $P(Y)$ or of $P(X)$ are kept fixed. The first provides the usual Shannon Coding Theorem, and the second a tuning theorem variant. It seems possible, however, that for many systems $P(X)$ and $P(Y)$ will interpenetrate. That is, $P(X)$ and $P(Y)$ will affect each other in characteristic ways, so that some form of mutual tuning may be the most effective strategy in a particular case.

1.2   The Rate Distortion Theorem

Suppose a sequence of signals is generated by a biological (or other) information source Y having output $y^n = y_1, y_2, \ldots$. This is 'digitized' in terms of the observed behavior of the system with which it communicates, for example a sequence of 'observed behaviors' $b^n = b_1, b_2, \ldots$. Assume each b^n is then deterministically retranslated back into a reproduction of the original biological signal, $b^n \rightarrow \hat{y}^n = \hat{y}_1, \hat{y}_2, \ldots$.

Define a distortion measure $d(y, \hat{y})$ comparing the original to the retranslated path. Many distortion measures are possible. The Hamming distortion is defined simply as $d(y, \hat{y}) = 1, y \neq \hat{y}, d(y, \hat{y}) = 0, y = \hat{y}$.

For continuous variates, the squared error distortion measure is just $d(y, \hat{y}) = (y - \hat{y})^2$.

The distortion between paths y^n and $\hat{y}^n$ is defined as $d(y^n, \hat{y}^n) \equiv \frac{1}{n} \sum_{j=1}^{n} d(y_j, \hat{y}_j)$.

A remarkable characteristic of the Rate Distortion Theorem is that the basic result is independent of the exact distortion measure chosen.

Suppose that with each path y^n and b^n-path retranslation into the y-language, denoted $\hat{y}^n$, there are associated individual, joint, and conditional probability distributions

$$p(y^n), p(\hat{y}^n), p(y^n, \hat{y}^n), p(y^n|\hat{y}^n).$$

The average distortion is defined as

$$D \equiv \sum_{y^n} p(y^n) d(y^n, \hat{y}^n) \tag{1.3}$$

It is possible to define the information transmitted from the Y to the $\hat{Y}$ process using the Shannon source uncertainty of the strings:

$$I(Y, \hat{Y}) \equiv H(Y) - H(Y|\hat{Y}) = H(Y) + H(\hat{Y}) - H(Y, \hat{Y}) \tag{1.4}$$

where $H(\ldots, \ldots)$ is the standard joint, and $H(\ldots|\ldots)$ the conditional, Shannon uncertainties.

If there is no uncertainty in Y given the retranslation $\hat{Y}$, then no information is lost, and the systems are in perfect synchrony.

In general, of course, this will not be true.

The rate distortion function $R(D)$ for a source Y with a distortion measure $d(y, \hat{y})$ is defined as

$$R(D) = \min_{p(y|\hat{y}); \sum_{(y, \hat{y})} p(y)p(y|\hat{y})d(y, \hat{y}) \leq D} I(Y, \hat{Y}) \tag{1.5}$$

The minimization is over all conditional distributions $p(y|\hat{y})$ for which the joint distribution $p(y, \hat{y}) = p(y)p(y|\hat{y})$ satisfies the average distortion constraint (i.e., average distortion $\leq D$).

The Rate Distortion Theorem states that $R(D)$ is the minimum necessary rate of information transmission which ensures the communication between the biological vesicles does not exceed average distortion D. Thus $R(D)$ defines a minimum necessary channel capacity. Cover and Thomas (2006) or Dembo and Zeitouni (1998) provide details. The rate distortion function has been calculated for a number of systems, often using Lagrange multiplier or Khun-Tucker optimization methods.

$R(D)$ is necessarily a decreasing convex function of D for any reasonable definition of distortion. That is, $R(D)$ is always a reverse J-shaped curve. This condition will prove central to characterizing the dynamics of systems that can be represented as having a RDF.

For the standard Gaussian channel having noise with zero mean and variance σ^2, using the squared distortion measure,

$$R(D) = 1/2\log[\sigma^2/D], 0 \leq D \leq \sigma^2$$

$$R(D) = 0, D > \sigma^2 \qquad (1.6)$$

Recall the relation between information source uncertainty and channel capacity: $H[X] \leq C$ where H is the uncertainty of the source X and C the channel capacity. Remember also that $C \equiv \max_{P(X)} I(X;Y)$, where $P(X)$ is chosen so as to maximize the rate of information transmission along a channel Y.

It is possible to give a rate distortion interpretation to the 'tuning theorem' version of the Shannon Coding Theorem. Shannon (1959) noted what he characterized as 'a curious and provocative duality' between the properties of an information source with a distortion measure and those of a channel. This duality is enhanced if we consider channels in which there is a cost associated with the different letters. Solving this problem corresponds to finding a source that is right for the channel and the desired cost. In a dual way, evaluating the rate distortion function for a source corresponds to finding a channel that is just right for the source and allowed distortion level. We will have more to say on this in the following chapter.

1.3   The Shannon-McMillan Theorem

The zero-error limit of the Rate Distortion Theorem is known as the Shannon-McMillan Theorem (SMT) (Cover and Thomas 2006), and serves as an important bridge to methods from statistical mechanics and nonequilibrium thermodynamics. The essential idea is that, for a broad class of information sources – stationary, ergodic – sufficiently long output sequences can be divided into two classes, the greatly larger being composed of 'nonsense' that does not conform to the 'grammar' and 'syntax' associated with the source. This set has, in the limit of long enough sequences, a vanishingly small probability of occurrence. The smaller set contains 'meaningful' sequences that do conform to grammar and syntax, and these are the high probability sequences. Let $N(n)$ be the number of meaningful sequences of length n emitted by an information source represented by the stochastic variate X. Then the SMT states that the uncertainty of the information source, $H[X]$, can be written as

$$
\begin{aligned}
H[X] &= \lim_{n \to \infty} \frac{\log[N(n)]}{n} \\
&= \lim_{n \to \infty} H(X_n | X_0, ..., X_{n-1}) \\
&= \lim_{n \to \infty} \frac{H(X_0, ..., X_n)}{n+1}
\end{aligned}
\tag{1.7}
$$

where $H(..|...)$ represents conditional and $H(.....)$ joint Shannon uncertainties.

In the limit of large n, $H[X]$ becomes homologous to the free energy density (FED) of a physical system at the thermodynamic limit of infinite volume. More explicitly, the FED of a physical system having volume V and partition function $Z(\beta)$ derived from a system's Hamiltonian, the energy function, at inverse temperature β (e.g., Landau and Lifshitz 2007) can be written as

$$
\begin{aligned}
F &= \lim_{V \to \infty} -\frac{1}{\beta} \frac{\log[Z(\beta, V)]}{V} \\
&\equiv \lim_{V \to \infty} \frac{\log[\hat{Z}(\beta, V)]}{V}
\end{aligned}
\tag{1.8}
$$

Feynman (2000), in fact, defines information as the free energy needed to erase a message, and shows how to construct an idealized machine that converts the information within a message into work, the definition of free energy.

In sum, information is a form of free energy and the construction and transmission of information within organisms consumes metabolic free energy, with great losses via the second law of thermodynamics. If there are limits on available metabolic free energy, there will necessarily be limits on the ability of the organism to properly process information.

The analogy between information and free energy can be extended.

1.4 Biological renormalizations

To summarize a long train of standard argument (Wilson 1971; Binney et al. 1986), imposition of invariance of H in equation (1.7) under a renormalization transform in some implicit 'size' parameter l leads to expectation of both a critical point in Q, the inverse available rate of metabolic free energy, written Q_C, reflecting a phase transition to or from collective behavior across an array of physiological subsystems interacting via information exchange, and of power laws for system behavior near Q_C. Addition of other parameters to the system results in a 'critical line' or surface.

Let $\kappa \equiv (Q_C - Q)/Q_C$ and take χ as the 'correlation length' defining the average domain in l-space for which the information source is primarily dominated by 'strong' ties that disjointly partition interacting subsystems. The first step is to average across l-space in terms of 'clumps' of length $L = < l >$. Then $H[J, Q, \mathbf{X}] \to H[J_L, Q_L, \mathbf{X}]$.

Taking Wilson's (1971) analysis as a starting point – not the only way to proceed – the 'renormalization relations' used here are:

$$H[Q_L, J_L, \mathbf{X}] = f(L)H[Q, J, \mathbf{X}]$$
$$\chi(Q_L, J_L) = \frac{\chi(Q, J)}{L} \tag{1.9}$$

with $f(1) = 1$ and $J_1 = J, Q_1 = Q$. The first equation significantly extends Wilson's treatment. It states that 'processing capacity,' as indexed by H, representing the 'richness' of the system, grows monotonically as $f(L)$, which must itself be a dimensionless function in L, since both $H[Q_L, J_L]$ and $H[Q, J]$ are themselves dimensionless. Most simply, this requires replacing L by L/L_0, where L_0 is the 'characteristic length' for the system over which renormalization procedures are reasonable, then setting $L_0 \equiv 1$, hence measuring length in units of L_0.

Wilson's original analysis focused on free energy density. Under 'clumping,' densities must remain the same, so that if $F[Q_L, J_L]$ is the free energy of the clumped system, and $F[Q, J]$ is the free energy density before clumping, then Wilson's (1971) equation (4) is $F[Q, J] = L^{-3}F[Q_L, J_L]$,

$$F[Q_L, J_L] = L^3 F[Q, J].$$

Remarkably, the renormalization equations are solvable for a broad class of functions $f(L)$, or more precisely, $f(L/L_0), L_0 \equiv 1$.

The second equation just states that the correlation length simply scales as L.

The central feature of renormalization in this context is the assumption that, at criticality, the system looks the same at all scales, that is, it is *invariant under renormalization* at the critical point. All else flows from this.

There is no unique renormalization procedure for information sources: other, very subtle, symmetry relations – not necessarily based on the elementary physical analog we use here – may well be possible. For example, McCauley (1993, p.168) describes the highly counterintuitive renormalizations needed to understand phase transition in simple 'chaotic' systems. This is important, since biological or social systems may well alter their renormalization properties – equivalent to tuning their phase transition dynamics – in response to external signals (Wallace 2005a).

To begin, following Wilson, take $f(L) = L^d$, d some real number $d > 0$, and restrict Q to near the 'critical value' Q_C. If $J \to 0$, a simple series expansion and some clever algebra gives

$$H = H_0 \kappa^\alpha$$
$$\chi = \frac{\chi_0}{\kappa^s} \tag{1.10}$$

where α, s are positive constants. More biologically relevant examples appear below.

Further from the critical point, matters are more complicated, appearing to involve Generalized Onsager Relations, 'dynamical groupoids', and a kind of nonequilibrium thermodynamics associated with a Legendre transform of H: $S \equiv H - QdH/dQ$.

An essential insight is that *regardless of the particular renormalization properties, sudden critical point transition is possible in the opposite direction for this model.* That is, going from a number of independent, isolated and fragmented systems operating individually and more or less at random, into a single large, interlocked, coherent structure, once the inverse index Q falls below threshold.

Thus, increasing free energy exchange or information crosstalk between them can bind several different cognitive 'language' functions into a single, embedding hierarchical metalanguage containing each as a linked subdialect, and do so in an inherently punctuated manner. This could be a dynamic process, creating a shifting, ever-changing pattern of linked submodules, according to the challenges or opportunities faced by the organism.

This heuristic insight can be made more exact.

Suppose that two ergodic information sources **Y** and **B** begin to interact, to 'talk' to each other, to influence each other in some way so that it is possible, for example, to look at the output of **B** – strings b – and infer something about the behavior of **Y** from it – strings y. We suppose it possible to define a retranslation from the B-language into the Y-language through a deterministic code book, and call $\hat{\mathbf{Y}}$ the translated information source, as mirrored by **B**.

Define some distortion measure comparing paths y to paths $\hat{y}$, $d(y, \hat{y})$. Invoke the Rate Distortion Theorem's mutual information $I(Y, \hat{Y})$, which is the splitting criterion between high and low probability pairs of paths. Impose, now, a parameterization by an inverse coupling strength Q, and a renormalization representing the global structure of the system coupling. This may be much different from the renormalization behavior of the individual components. If $Q < Q_C$, where Q_C is a critical point (or surface), the two information sources will be closely coupled enough to be characterized as condensed.

In the absence of a distortion measure, the Joint Asymptotic Equipartition Theorem gives a similar result.

Detailed coupling mechanisms will be sharply constrained through regularities of grammar and syntax imposed by limit theorems associated with phase transition.

Next the mathematical detail concealed by the invocation of the asymptotic limit theorems of information theory emerges with a vengeance. Equation (1.9) states that H and the correlation length, the degree of coherence on the underlying network, scale under renormalization clustering in chunks of size L as

$$H[Q_L, J_L]/f(L) = H[J, K]$$

$$\chi[Q_L, J_L]L = \chi(K, J),$$

with $f(1) = 1, Q_1 = Q, J_1 = J$, where we have rearranged terms.

Differentiating these two equations with respect to L, so that the right hand sides are zero, and solving for dQ_L/dL and dJ_L/dL gives, after some consolidation,

$$dQ_L/dL = u_1 d\log(f)/dL + u_2/L$$
$$dJ_L/dL = v_1 J_L d\log(f)/dL + \frac{v_2}{L}J_L \tag{1.11}$$

The $u_i, v_i, i = 1, 2$ are functions of Q_L, J_L, but not explicitly of L itself. Expand these equations about the critical value $Q_L = Q_C$ and about $J_L = 0$,

$$dQ_L/dL = (Q_L - Q_C)y d\log(f)/dL + (Q_L - Q_C)z/R$$
$$dJ_L/dL = w J_L d\log(f)/dL + x J_L/L \tag{1.12}$$

The terms $y = du_1/dQ_L|_{Q_L=Q_C}, z = du_2/dQ_L|_{Q_L=Q_C}, w = v_1(Q_C, 0), x = v_2(Q_C, 0)$ are constants.

Solving the first of these equations gives

$$Q_L = Q_C + (Q - Q_C)L^z f(L)^y \tag{1.13}$$

again remembering that $Q_1 = Q, J_1 = J, f(1) = 1$.

Wilson (1971) iterates this relation, which is supposed to converge rapidly near the critical point, assuming that for Q_L near Q_C,

$$Q_C/2 \approx Q_C + (Q - Q_C)L^z f(L)^y \tag{1.14}$$

Now iterate in two steps, first solving this for $f(L)$ in terms of known values, and then solving for L, finding a value L_C that we then substitute into the first of equations (1.9) to obtain an expression for $H[Q, 0]$ in terms of known functions and parameter values.

The first step gives the general result

$$f(L_C) \approx \frac{[Q_C/(Q_C - Q)]^{1/y}}{2^{1/y} L_C^{z/y}} \tag{1.15}$$

Solving this for L_C and substituting into the first expression of equation (1.9) gives, as a first iteration of a far more general procedure (Shirkov and Kovalev 2001), the result

$$H[Q, 0] \approx \frac{H[Q_C/2, 0]}{f(L_C)} = \frac{H_0}{f(L_C)}$$
$$\chi(K, 0) \approx \chi(Q_C/2, 0)L_C = \chi_0 L_C \tag{1.16}$$

which are the essential relationships.

Note that a power law of the form $f(L) = L^m, m = 3$, which is the direct physical analog, may not be biologically reasonable, since it says that 'language richness' can grow very rapidly as a function of increased network size. Such rapid growth is simply not observed.

Taking the biologically realistic example of non-integral 'fractal' exponential growth,

$$f(L) = L^\delta \qquad (1.17)$$

where $\delta > 0$ is a real number which may be quite small, equation (1.15) can be solved for L_C, obtaining

$$L_C = \frac{[Q_C/(Q_C - Q)]^{[1/(\delta y+z)]}}{2^{1/(\delta y+z)}} \qquad (1.18)$$

for Q near Q_C. Note that, for a given value of y, one might characterize the relation $\alpha \equiv \delta y + z = $ constant as a 'tunable universality class relation' in the sense of Albert and Barabasi (2002).

Substituting this value for L_C back into equation (1.15) gives a complex expression for H, having three parameters: δ, y, z.

A more biologically relevant form of $f(L)$ is a logarithmic curve that 'tops out',

$$f(L) = m \log(L) + 1 \qquad (1.19)$$

Again $f(1) = 1$.

Using Mathematica 4.2 or above to solve equation (1.15) for L_C gives

$$L_C = [\frac{A}{LambertW[A \exp(z/my)]}]^{y/z} \qquad (1.20)$$

where $A \equiv (z/my)2^{-1/y}[Q_C/(Q_C - Q)]^{1/y}$.

The transcendental function LambertW(x) is defined by the relation

$$LambertW(x) \exp(LambertW(x)) = x.$$

It arises in the theory of random networks and in renormalization strategies for quantum field theories.

An asymptotic relation for $f(L)$ would be of particular biological interest, implying that 'language richness' increases to a limiting value with population growth.

Taking

$$f(L) = \exp[m(L - 1)/L] \qquad (1.21)$$

gives a system which begins at 1 when $L = 1$, and approaches the asymptotic limit $\exp(m)$ as $L \to \infty$. Mathematica finds

$$L_C = \frac{my/z}{LambertW[B]}, \qquad (1.22)$$

where

$$B \equiv (my/z) \exp(my/z)[2^{1/y}[Q_C/(Q_C - Q)]^{-1/y}]^{y/z}$$

These developments take the theory significantly beyond arguments by abduction from simple physical models.

Note that, while we have focused on a single information source H, it is possible to expand the argument to the renormalization symmetries of a Morse Function constructed from a set of information source uncertainties, as will be done below.

1.5   The Data Rate Theorem

The Data Rate Theorem (DRT), a generalization of the classic Bode Integral Theorem for linear control systems, bridges a longstanding gap between information theory and control theory. It describes the stability of feedback control under data rate constraints (Nair et al. 2007). Given a noise-free data link between a discrete linear plant and its controller, unstable modes can be stabilized only if the feedback data rate $\mathcal{I}$ is greater than the rate of 'topological information' generated by the unstable system. For the simplest incarnation, if the linear matrix equation of the plant is of the form $x_{t+1} = \mathbf{A}x_t + ...$, where x_t is the n-dimensional state vector at time t, then the necessary condition for stabilizability is that

$$\mathcal{I} > \log[|det\mathbf{A}^u|] \tag{1.23}$$

where det is the determinant and $\mathbf{A}^u$ is the decoupled unstable component of $\mathbf{A}$, i.e., the part having eigenvalues ≥ 1. The determinant represents a generalized volume. Thus there is a critical positive data rate below which there does not exist any quantization and control scheme able to stabilize an unstable system (Nair et al. 2007).

The new theorem, and its variations, relate control theory to information theory and are as fundamental as the Shannon Coding and Source Coding Theorems, and the Rate Distortion Theorem for understanding complex cognitive machines and biological phenomena.

The Data Rate Theorem can be reframed in terms of metabolic free energy as providing the essential control signal, an argument that will recur frequently, explicitly and implicitly.

Suppose an intensity of available metabolic or other free energy is associated with joint and individual information sources having Shannon uncertainties $H(X,Y), H(X), H(Y)$, e.g., rates $\mathcal{H}_{X,Y}$, $\mathcal{H}_X, \mathcal{H}_Y$.

Although, as Feynman (2000) argues, information is a form of free energy, there is necessarily a massive entropic loss in its actual expression, so that a probability distribution of a source uncertainty H might be written in Gibbs form as

$$P[H] \approx \frac{\exp[-H/\omega \mathcal{H}]}{\int \exp[-H/\omega \mathcal{H}]dH} \tag{1.24}$$

assuming ω to be very small.

To first order

$$\hat{H} \equiv \int H P[H]dH \approx \omega \mathcal{H} \tag{1.25}$$

and, using the well-known relation (Cover and Thomas 2006)

$$H(X,Y) \leq H(X) + H(Y) \tag{1.26}$$

then

$$\hat{H}(X,Y) \leq \hat{H}(X) + \hat{H}(Y)$$
$$\mathcal{H}_{X,Y} \leq \mathcal{H}_X + \mathcal{H}_Y \tag{1.27}$$

Thus, allowing crosstalk between information sources consumes a lower rate of free energy than isolating them: it takes more free energy to isolate information sources than allowing them to engage in crosstalk.

Conversely, at the free energy expense of supporting two information sources, $- X$ and Y together $-$ it is possible to catalyze a set of joint paths defined by their joint information source. In consequence, given a physiological or other system (or set of them) having an associated information source $H(...)$, an external information source Y can catalyze the joint paths associated with the joint information source $H(...,Y)$ so that a particular chosen reaction path has the lowest relative free energy.

In sum, at the expense of larger global free information expenditure – maintaining two (or more) information sources with their entropic losses instead of one – the system can support the generalized physiology of a Maxwell's Demon, doing work so that regulatory signals can direct system response, thus locally reducing uncertainty at the expense of larger global entropy production. The free energy feeding the regulatory information source Y acts as the feedback control signal in a DRT model.

Chapter 2

A symmetry-breaking model

2.1 Summary

The rate of metabolic free energy availability serves as a temperature analog for the 'spontaneous symmetry breaking' of the group structure associated with the error minimization coding scheme of protein folding, characterizing a phase transition that collapses normal folding to a generalized autocatalytic amyloid production in Alzheimer's disease. Prion and intrinsically disordered protein (IDP) pathologies may present certain parallels, and extension of the argument is possible to the tiling symmetries that are central to the pathologies of more complex biological processes, taking Maturana's perspective regarding the central role of cognition at every scale and level of organization of the living state.

2.2 Introduction

As described, Swerdlow's mitochondrial cascade hypothesis for Alzheimer's disease (AD) (Swerdlow et al. 2010) proposes that a person's genes determine their baseline mitochondrial function and durability. While both parents influence one's lifetime AD risk, since mtDNA is maternally inherited, mothers have a greater impact than fathers. Mitochondrial function declines with age, and this drives a variety of age-associated physiological changes. Cell physiology may initially compensate for and adapt to this change, but eventually adequate compensation is no longer possible. The mitochondrial cascade hypothesis proposes that a genetically determined mitochondrial starting line, in conjunction with a genetically and environmentally determined rate of mitochondrial decline, determines the age at which clinical disease ensues.

Here, we will propose a strikingly direct model of this dynamic that generalizes across much of the phenotype and pathology of development and aging, may apply to some pathologies of high order cognition (Scaglia 2010), and seems relevant to certain other mitochondrial diseases. Indeed, as is well known, aging in general is closely linked to cellular mitochondrial function (e.g., D.C. Wallace 2005, 2010). As Lee and Wei (2012) put it, aging is a degenerative process that is associated with progressive accumulation of deleterious changes with time, reduction of physiological function and increase in the chance of disease and death. Studies reveal a wide spectrum of alterations in mitochondria and mitochondrial DNA [mtDNA] with aging. Mitochondria are the main cellular energy sources that generate the cellular energy source ATP through respiration and oxidative phosphorylation in the inner membrane of mitochondria. The respiratory chain of that system is also the primary intracellular source of reactive oxygen species and free radicals under normal physiological and pathological conditions, so that mitochondria play a central role in a great variety of cellular processes.

Many biochemical studies of isolated mitochondria reveal that the electron transport activities of respiratory enzyme complexes gradually decline with age in the brain, skeletal muscle, liver and skin fibroblasts of normal human subjects. Numerous molecular studies demonstrated that somatic mutations in mitochondrial DNA accumulate with age in a variety of tissues in humans. These age-associated changes in mitochondria are well correlated with the deteriorative processes of tissues in aging.

However, although abundant experimental data now support the concept that decline in mitochondrial energy metabolism, reactive oxygen species overproduction and accumulation of mtDNA mutations in tissue cells are important contributors to human aging, the detailed mechanisms by which these biochemical events cause aging have remained to be established.

Similarly, Park and Larsson (2011) conclude

> The different types of pathologies that are caused by mtDNA mutations are remarkable, and in many cases there is a reasonably good correlation between genotype and phenotype. However, the underlying pathophysiology is not understood in any depth. Important challenges for the future involve understanding the downstream effects of mitochondrial dysfunction on cell physiology in disease and aging.

More generally, mitochondrial dysfunction is increasingly implicated in broadly developmental diseases such as autism and schizophrenia (e.g., Scaglia 2010; Shao et al. 2008; Clay et al. 2011).

To reach an understanding of downstream effects, we begin far afield indeed.

2.3 Biological symmetries

Critical phenomena in nonequilibrium biological processes have been the subject of intense study for some time. Smith et al. (2011), for example, examine biomolecular signal transduction switching using cutting-edge techniques including operator and functional integral methods from reaction-diffusion theory, finding that

> The lack of convenient symmetries in real biomolecular systems promises to make analysis intractable for most quantitative phenomenology, and a recourse to numerics is likely to be the only general-purpose solution.

In reality, the symmetries are there, but uncovering them is not straightforward, and many are more akin to Arabic tilings than simple rotations or translations. Indeed, a remarkable, but seemingly under appreciated, theoretical development has been the finding of a close relation between information theory inequalities and a spectrum of results in the theory of finite groups (e.g., Yeung 2008):

Given two random variables X_1 and X_2 having Shannon uncertainties $H(X_1)$ and $H(X_2)$ defined in the usual manner, the information theory chain rule states that, for the joint uncertainty $H(X_1, X_2)$,

$$H(X_1) + H(X_2) \geq H(X_1, X_2) \tag{2.1}$$

Similarly, let G be any finite group, and G_1, G_2 be subgroups of G. Let $|G|$ represent the order of a group, i.e., the number of elements. Then it is easy to show the intersection $G_1 \cap G_2$ is also a subgroup, and that

$$\log[\frac{|G|}{|G_1|}] + \log[\frac{|G|}{|G_2|}] \geq \log[\frac{|G|}{|G_1 \cap G_2|}] \tag{2.2}$$

Defining a probability for a 'random variate' associated with a group G as $Pr\{X = a\} = 1/|G|$ permits construction of a group-characterized information source, noting that, in general, the joint uncertainty of a set of random variables in not necessarily the logarithm of a rational number. The surprising ultimate result, however, is that there is a one-to-one correspondence between unconstrained information inequalities and group inequalities. Indeed, unconstrained inequalities can be proved by techniques in group theory, and certain group-theoretic inequalities can be proven by techniques of information theory.

More generally, the theory of error-correcting codes, usually called algebraic coding theory (Pretzel 1996, Roman 1997; van Lint 1999), seeks particular redundancies in message coding over noisy channels that enable efficient reconstruction of lost or distorted information. The full-bore panoply of groups, ideals, rings, algebras, and finite fields is brought to bear on the problem to produce a spectrum of codes having different capabilities and complexities: BCH, Goppa, Hamming, Linear, Reed-Muller, Reed-Solomon, and so on.

Here, we will provide examples suggesting that the relations between groups, groupoids, and a broad spectrum of information related phenomena of interest in biology are, similarly, surprisingly intimate.

Group symmetries associated with an error-minimization coding scheme – as opposed to error correction coding – will dominate a necessary conditions statistical model of a 'spontaneous symmetry breaking' phase transition that drives the collapse of protein folding to pathological amyloid production, and groupoids emerge as central in the study of a similar wide-ranging 'ground state' failure of cognitive process, adopting the Maturana/Varela (Maturana and Varela 1980) perspective on the necessity of cognition at every scale and level of organization of the living state.

2.4   Group structure of biological codes

Tlusty's (2007) analysis of deterministic error-limiting codes (DEL) that minimize the impact of coding errors provides a basis for examining the problem of amyloid protein misfolding. Tlusty (2008) models the emergence of the genetic code as a transition in a noisy information channel, using a Rate Distortion Theorem methodology. After some development, Tlusty finds the number of possible amino acids in a coding scheme is analogous to the well-known topological coloring problem. But while in the coding problem one desires maximal similarity in the colors of neighboring 'countries', in the coloring problem one must color neighboring countries by different colors. Explicitly, one uses Heawood's formula (Ringel and Young

1968) to determine the number of possible 'amino acids' given a codon graph designed to minimize errors in coding:

$$chr(\gamma) = Int[\frac{1}{2}(7 + \sqrt{1 + 48\gamma})] \qquad (2.3)$$

where $chr(\gamma)$ is the number of 'colored' regions, Int is the integer value of the enclosed expression, and γ is the genus of the surface of the underlying code network – basically the number of 'holes' in the code network. In general, $\gamma = 1 - (1/2)(V - E + F)$, where V is the number of code network vertices, E the number of network edges, and F the number of enclosed faces.

The central trick is that one can obtain, for any DEL code, a basic group theoretic characterization by noting that the fundamental group (FG) of a closed, orientable surface of genus γ – in which the code network is taken as embedded – is the quotient of the free group on the 2γ generators $a_1, ..., a_\gamma, b_1, ..., b_\gamma$ by the normal subgroup generated by the product of the commutators

$$a_1 b_1 a_1^{-1} b_1^{-1} ... a_\gamma b_\gamma a_\gamma^{-1} b_\gamma^{-1} \qquad (2.4)$$

This is a standard construction (e.g., Lee 2000). For example, the FG of a sphere, an orientable surface with zero holes, is trivial, having only one element, while that of the torus – a donut-like orientable surface with one hole – is isomorphic to the direct product of the integers, written as $\mathcal{Z} \times \mathcal{Z}$, and so on.

That is, every DEL biological code is associated with a fundamental group. The more complex the code, the richer the symmetries of the associated error network, seen as embedded in a smooth surface of genus γ. Indeed, a weakened 'groupoid' version of the argument will prove central to understanding the structure of cognitive process, as developed in a following section.

Wallace (2010) suggests that the overall scheme applies to a 'protein folding code' as well. Hecht et al. (2004) note that protein α-helices have the underlying 'code' 101100100110... where 1 indicates a polar and 0 a non-polar amino acid. Protein β-sheets, by contrast, have the simpler basic 'code' 10101010...

Equation (2.3), most directly, produces the table

γ (# surface holes)	chr(γ) (# error classes)
0	4
1	7
2	8
3	9
4	10
5	11
6, 7	12
8, 9	13

In Tlusty's scheme, the second column represents the maximal possible number of product classes that can be reliably produced by error-prone codes having γ holes in the underlying coding error network.

Normal irregular protein symmetries were first classified by Levitt and Chothia (1976), following a visual study of polypeptide chain topologies in a limited dataset of globular proteins. Four major classes emerged; all α-helices; all β-sheets; α/β; and $\alpha + \beta$, with the latter two having the obvious meaning.

While this scheme strongly dominates observed irregular protein forms, Chou and Maggiora (1998), using a much larger data set, recognize three more 'minor' symmetry equivalence classes; μ (multi-domain); σ (small protein); and ρ (peptide), and a possible three more subminor groupings.

We infer that, from Tlusty's perspective, the normal globular 'protein folding code error network' is, essentially, a large connected 'sphere' – producing the four dominant structural modes – but having as many as three more attachment handles, in the Morse Theory sense (Matusumoto 2001). These basic entities then act to produce an almost unlimited set of functional proteins under normal conditions.

2.5   The amyloid condensation

What happens to the fundamental group of a deterministic error-limiting code under conditions that are not normal? A clue can be found in speculations about a supposed prebiotic 'amyloid world' as postulated by Maury (2009). This, in contrast to the current rich variety of normal protein structures and functions, would have been built on a single β-sheet lamination, and could show, by contrast to the normal protein world, at best a starkly simple structural spectrum based on replications of an eight-fold steric zipper (Sawaya et al. 2007).

As Goldschmidt et al. (2010) put the matter, regarding what they characterized as the 'amylome',

> We found that [protein segments with high fibrillation propensity] tend to be buried or twisted into unfavorable conformations for forming beta sheets... For some proteins a delicate balance between protein folding and misfolding exists that can be tipped by changes in environment, destabilizing mutations, or even protein concentration...
>
> In addition to... self-chaperoning effects... proteins are also protected from fibrillation during the process of folding by molecular chaperones...
>
> Our genome-wide analysis revealed that self complementary segments are found in almost all proteins, yet not all proteins are amyloids. The implication is that chaperoning effects have evolved to constrain self-complementary segments from interaction with each other.

Gamerdinger and Deuerling (2014) elaborate these ideas:

> Molecular chaperones are found in all cells and are essential for maintaining a functional proteome. The main function of chaperones is to promote correct protein folding by protecting non-native proteins from folding along pathways that lead to protein misfolding and aggregation. To fulfill this task, chaperones must recognize a non-native protein, transiently bind to it, and then release it at precisely the right time to allow the substrate to proceed with its folding course. Many but not all chaperones use adonosine $5'$-triphosphate (ATP) [the central metabolic free energy source] to control the dynamic substrate binding and release cycle [Kim et al., 2013].

As Yao et al. (2009) put the matter,

> Alzheimer's pathology is accompanied by a decrease in expression and activity of enzymes involved in mitochondrial bioenergetics... [T]here is a generalized shift from glycolytic energy production toward use of an alternative fuel, keytone bodies. This is evidenced by a 45% reduction in

cerebral glucose utilization in AD patients... Patients with incipent AD exhibit a utilization ratio of 2:1 glucose to alternative fuel, whereas comparably aged controls exhibit a ratio of 29:1... [In comparison,] young controls exclusively use glucose...

Absent adequate rates of available metabolic free energy – delivered through ATP or other means (Saio et al. 2014) the 'fundamental group' structure of the protein folding system essentially disappears in a Landau-like spontaneous symmetry breaking leading to amyloid formation, which may trigger a catastrophic autocatalytic dynamic. As Querfurth and LaFerla (2010) put it

> $A\beta$ is a potent mitochondrial poison, especially affecting the synaptic pool.. In Alzheimer's disease, exposure to $A\beta$ inhibits key mitochondrial enzymes in the brain and in isolated mitochondria... Cytochrome c oxidase is specifically attacked.. Consequently, electron transport, ATP production, oxygen consumption, and mitochondrial membrane potential all become impaired... [I]nstability and the irreparability of the brain's mitochondrial genome allow the gradual accumulation of [deleterious] mtDNA mutations...

Effective chaperoning requires considerable metabolic energy, and failure to provide levels adequate for both maintaining and operating such biochemical translation machinery triggers a canonical 'code collapse'. This is most likely to take place in a highly punctuated manner, leading to a complex, two-fold autocatalytic runaway pathology in which amyloid plaques both catalyze their own growth (e.g., Sabate et al. 2003) and damage the chaperone mechanisms designed to overcome such protein misfolding.

We choose a classic formalism to examine the first stage of this dynamic, but other approaches to constructing an appropriate Morse Function (Pettini 2007) are possible.

The existence of a Tlusty-like error minimization coding structure implies the existence of some information source using that code-and-translator or code-and-chaperone channel. As Feynman (2000), following Bennett (1988) argues, it is possible to make a small (idealized) machine that transforms information received into work – free energy. Indeed, Feynman defines information precisely in terms of the free energy needed to erase it. Representing the intensity of available mitochondrial free energy as $\mathcal{H}$, we write a pseudoprobability for an information source X_j associated with coding mode j and having source uncertainty H_j as

$$\Pr[H_j] = \frac{\exp[-H_j/\omega\mathcal{H}]}{\sum_{i=1}^{n}\exp[-H_i/\omega\mathcal{H})]} \tag{2.5}$$

where ω is an appropriate scaling constant.

This leads to construction of a 'free energy' Morse Function, F, defined in terms of the rate of available metabolic free energy as

$$\exp[-F/\omega\mathcal{H}] = \sum_{i=1}^{n}\exp[-H_i/\omega\mathcal{H}] \tag{2.6}$$

See the Mathematical Appendix for a summary of standard material on Morse Functions.

The central insight regarding phase transitions in physical systems is that certain critical phenomena take place in the context of a significant alteration in symmetry, with one phase being far more symmetric than the other (Landau and Lifshitz 2007; Pettini 2007). A symmetry is lost in the transition – spontaneous symmetry breaking. The greatest possible set of symmetries in a physical system is that of the Hamiltonian describing its energy states. Usually states accessible at lower temperatures will lack the symmetries available at higher temperatures, so that the lower temperature phase is less symmetric. The randomization of higher temperatures ensures that higher symmetry/energy states will then be accessible to the system. The shift between symmetries is highly punctuated in the temperature index.

This line of argument suggests the existence of complex forms of highly punctuated phase transition in code/translator function with changes in demand for, or supply of, the rate of metabolic free energy needed to run the protein chaperone machine. That is, applying a spontaneous symmetry breaking argument to the Morse Function F generates topological transitions involving changes in the fundamental group defined by error code graph structure as the mitochondrial 'temperature' $\mathcal{H}$ decreases. As the rate of delivery of the free energy running the chaperone machines decreases, complex coding schemes can no longer be sustained, driving a punctuated shift of the fundamental group of the protein folding code to a degenerate, collapsed amyloid state.

Details of such an information phase transition may also be described using 'biological' renormalization methods of Chapter 1 that are analogous to, but much different from, those used in the determination of physical phase transition universality classes (Wilson 1971). Suppose, in manner of Chapter 1, it is possible to define a characteristic 'length', say l, on the system. It is then again possible to define renormalization symmetries in terms of the 'clumping' transformation on F, so that, for clumps of size L, in an external 'field' of strength J (that can be set to 0 in the limit), one can write, in the usual manner

$$F[Q(L), J(L)] = f(L)F[Q(1), J(1)]$$
$$\chi(Q(L), J(L)) = \frac{\chi(Q(1), J(1))}{L} \tag{2.7}$$

where χ is a characteristic correlation length and Q is an 'inverse temperature measure', i.e., $\propto 1/\omega\mathcal{H}$.

As described in Chapter 1, very many 'biological' renormalizations, $f(L)$, are possible that lead to a number of quite different universality classes for biological phase transition. Indeed, a 'universality class tuning' can be used as a tool for large-scale regulation of the system. Again, while Wilson (1971) necessarily uses $f(L) \propto L^3$ for simple physical systems, following Wallace (2005a), it is possible to argue that, since F is so closely related to information measures, it is likely to 'top out' at different rates with increasing system size, so other forms of $f(L)$ must be explored. Indeed, standard renormalization calculations for $f(L) \propto L^\delta, m\log(L) + 1$, and $\exp[m(L-1)/L]$ all carry through.

Matters may, in fact, be even more complicated. Kim and Hecht (2006) suggest that overall amyloid fibril geometry is very much driven by the underlying β-sheet coding 1010101, although the rate of fibril formation may be determined by exact chemical constitution. Sawaya et al. (2007) parse some of those subtleties. As described, they identify an eight-fold 'steric zipper' symmetry necessarily associated with the linear amyloid fibrils. In essence, two identical sheets can be classified by the orientation of their faces (face-to-face/face-to-back), the orientation of their strands (with both sheets having the same edge of the strand up or one up and the other down), and whether the strands within the sheets are parallel or anti parallel. Five of the eight symmetry possibilities have been observed.

As mentioned earlier, Maury's (2009) 'amyloid world' model for the emergence of prebiotic informational entities, based on the extraordinary stability of amyloid structures in the face of the harsh conditions of the prebiotic world, provides some further insights. From this perspective, the synthesis of RNA, and the evolution of the RNA-protein world, were later, but necessary events for further bimolecular evolution. Maury further argues that, in the contemporary DNA⇔RNA⇒protein world, the primordial β-conformation-based information system is preserved in the form of a cytoplasmic epigenetic memory.

Falsig et al. (2008) examine the many different strains of prions, finding that differences in kinetics of the elementary steps of prion growth underlie the differential proliferation of prion strains, based on differential frangibility of prion fibrils. They argue that an important factor is the size of the stabilizing cross-β amyloid core that appears to define the physical properties of the resulting structures, including their propensity to fragment, with small core sizes leading to enhanced frangibility. In terms of the protein folding funnel approach, they find that intrinsic frustration implies that several distinct arrangements favoring a certain subset of globally incompatible interactions are possible, reflecting the observed strain-dependent differences in the parts of the sequence incorporated into the fibril core.

In addition, they argue, there are unexplored similarities between Alzheimer's and prion diseases, that is, the analogies between prion and Aβ aggregates could be broader than initially suspected. Recent work on the transmissibility of different strains of Alzheimer's disease in mouse models seems to definitively confirm those speculations (Watts et al. 2014; Stohr et al. 2014; Aguzzi 2014). Earlier epidemiological studies identified different forms of AD that may be associated with different autocatalytic amyloid subtypes (Mayeux et al. 1985; Friedland et al. 1988; Richie and Touchon 1992).

As Westermark (2005) puts it,

> Amyloid fibril proteins vary greatly in amino acid sequence and size, yet amyloid fibrils are similar in morphological appearance and in many properties. In protein solutions above the critical concentration, fibrils form spontaneously after a lag phase, which may vary considerably. Seeding a solution of an amyloidgenic peptide with preformed fibrils reduces the lag phase significantly by the elimination of nucleation. In seeding, it is believed that the seed grows by elongation at the ends of the fibrils where new monomers or preformed smaller aggregates are added. Interestingly, *in vivo* one and the same peptide can obtain different morphologies, depending on the seed. Thus the daughter fibrils tend to have the same morphology as the

mother... Mixed amyloids seem... to be very uncommon.

Following Wallace and Wallace (2011), given the eight-fold symmetry possible to the amyloid fiber, say versions A → H, then the simplest 'amyloid frangibility code' is the set of identical pairings

$$\{AA, BB, ..., GG, HH\}$$

producing eight different possible structures and their reproduction by fragmentation-and-growth, analogous to, but simpler than, double helix reproduction. More complex symmetries, or the possibility of combinatorial recombinations, would allow a much richer structure, producing distinct Alzheimer's disease species or quasi-species, in the sense of Collinge and Clarke (2007). Permitting different sequence lengths or explicitly identifying different sequence orders would vastly enlarge what Collinge has characterized as a 'cloud' of possibilities in the case of prion diseases, but likely applicable as well to Alzheimer's disease(s), although, as Westermark (2005) comments, mixed amyloids seem relatively rare.

Recent work on prions appears to support something of Maury's hypothesis. Li et al. (2010) find that infectious prions, mainly what is called PrP^{Sc}, a spectrum of β sheet-rich conformers of the normal host protein PrP^C, undergo Darwinian evolution in cell culture. In that work, prions show the evolutionary hallmarks: they are subject to mutation, as evidenced by heritable changes of their phenotypes, and to selective amplification, as found by the emergence of distinct populations in different environments. One might speculate that prion diseases and the different strains of Alzheimer's disease represent fossilized remains of Maury's prebiotic amyloid world.

A principal outcome of a full analysis, however, is that amyloid-β disease is not a simple matter of self-driven autocatalytic reaction dynamics, as Sabate et al. (2003) and many others postulate. To paraphrase Querfurth and LaFerla (2010), there is a heterogeneity of pathways that could initiate and drive AD. There is no single linear chain of events. Indeed, some changes are not pathologic but reactionary or protective. Thus, a multitargeted approach to prevent or treat AD, as used for other multigenic disorders, is needed. They conclude that many current approaches, including the 'amyloid hypothesis', may be minor or wrong, and it may be that some critical aging-related process is the disease trigger.

In sum, AD is a complicated gestalt phenomenon that involves multifactorial deterioration in, or overwhelming of, the sophisticated cognitive biological control structures associated with chaperone dynamics.

2.6 Biomolecular generalizations

Reconsidering prion disease

As Cobb et al. (2007) put it, transmissible spongiform encephalopathies (TSE) represent a group of fatal neurodegenerative diseases associated with the conversion of the normally monomeric and α-helical protein PrP^C, to the β-sheet-rich PrP^{Sc} form, believed to be the main component of the infective agent. See figure 2.1.

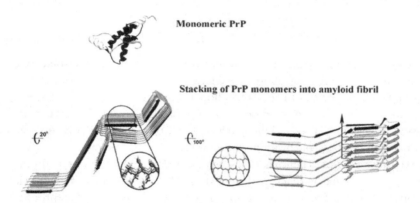

Fig. 2.1 Adapted from Cobb et al. 2007. Monomeric PrP^C is an assembly of α-helicies, stabilized and made physiologically active through interaction with a cellular lipid membrane. The pathological PrP^{Sc} form represents an amyloid condensation. The central point is that the PrP^C/membrane composite can be viewed as a kind of inherent chaperone system whose failure, triggered by an infectious agent into autocatalytic catastrophe, overcomes available rates of mitochondrial-delivered metabolic free energy.

While the amyloid conformation has the lowest chemical free energy, the normal form is anchored to a lipid membrane, whose total assembly has a relatively lower free energy (e.g., Riesner, 2003). This can be modeled, in our formalism, by means of the information theory chain rule of equation (2.1). Again, we assume an underlying 'code' linked with an information source that uses it.

Recapitulating something of the argument of Section 1.5, suppose an intensity of available free energy is associated with each defined joint and individual information source having Shannon uncertainties $H(X,Y), H(X), H(Y)$, e.g., rates $\mathcal{H}_{X,Y}$, $\mathcal{H}_X$,$\mathcal{H}_Y$.

Although information is a form of free energy, there is necessarily a massive entropic loss in its actual expression, so that the probability distribution of a source uncertainty H might again be written in Gibbs form as

$$P[H] \approx \frac{\exp[-H/\omega\mathcal{H}]}{\int \exp[-H/\omega\mathcal{H}]dH} \tag{2.8}$$

assuming ω is very small.

To first order, then,

$$\hat{H} \equiv \int H P[H]dH \approx \omega\mathcal{H} \tag{2.9}$$

and, using equation (2.1),

$$\hat{H}(X,Y) \leq \hat{H}(X) + \hat{H}(Y)$$
$$\mathcal{H}_{X,Y} \leq \mathcal{H}_X + \mathcal{H}_Y \tag{2.10}$$

Allowing crosstalk consumes a lower rate of free energy than isolating information sources; it takes more free energy – higher total cost – to insulate information sources than it does to allow them to engage in crosstalk.

Thus, at the free energy expense of supporting X and Y together, it is possible to catalyze a set of joint paths defined by their joint source so that a particular reaction pathway has the lowest relative free energy.

We must assume, then, that the system of PrP^C/membrane is, in fact, a composite regulator-reactor, designed, at the expense of mitochondrial-driven metabolic free energy, to both use PrP^C for cellular processes and to prevent pathological amyloid condensation. Thus the question is not the change in 'symmetry' of the molecular transition $PrP^C \rightarrow PrP^{Sc}$, but rather the change in 'functional symmetry' of the transition PrP^C/membrane $\rightarrow PrP^{Sc}$, which is a different matter altogether (e.g., Laganowsky et al. 2014). That latter transition, in the case of a TSE, appears catalyzed by an infectious agent that can overwhelm stabilizing or corrective mechanisms.

That is, it now seems necessary to extend the model to a spectrum of biological automata beyond protein chaperones. This, we will argue, suggests in turn that mitochondrial dysfunction provides a good part of the underlying trigger mechanism Querfurth and LaFerla (2010) suggest.

Intrinsically disordered proteins

As Csizmok and Tompa (2009) indicate, intrinsically disordered proteins (IDP) lack a well-defined structure, yet carry out important functions often associated with the regulation of cell cycle and transcription, and their mutations are frequently involved in neurodegenerative diseases, as usual, caused by the structural transition of disordered proteins to insoluble, highly ordered amyloid deposits.

Wallace (2012b) applies nonrigid molecule formalism to IDP, leading to very complex symmetry states for functional IDP's that collapse upon amyloid condensation. The argument is worth summarizing.

Longuet-Higgins (1963), in his classic paper, argues that the symmetry group of a nonrigid molecule is the set of all feasible permutations of the positions and spins of identical nuclei and of all feasible permutation-inversions, which simultaneously invert the coordinates of all particles in the center of mass.

The theory arising from this insight has had great success for understanding the spectra of modestly large molecules across much of chemistry and chemical physics. It can, with some development, be applied to the problem of understanding normal IDP dynamics.

Assume it possible to extend nonrigid molecular group theory to the long, whip-like frond of an IDP anchored at both ends, via a sufficient number of semidirect and/or wreath products over an appropriate set of finite and/or compact groups (e.g., Balasubramanian 1980). These are taken as parameterized by an index of 'frond length' L which might simply be the total number of amino acids in the IDP. In general, the number of group elements can be expected to grow exponentially, as $\sum \Pi_j |G_j||H_j|^L$, where $|G_k|$ and $|H_k|$ are the size, in an appropriate sense, of symmetry groups G_k and H_k. Hence, for large L, we are driven to a spontaneous symmetry breaking statistical mechanics approach on a Morse function, following the arguments of Pettini (2007). Typically, many such Morse functions are possible, and we construct one using group representations. See the Mathematical Appendix for a brief summary of standard material on Morse Theory.

Take an appropriate group representation by matrices and construct a pseudo probability $\mathcal{P}$ for nonrigid group element ω as

$$\mathcal{P}[\omega] = \frac{\exp[-|\chi_\omega|/\kappa L]}{\sum_\nu \exp[-|\chi_\nu|/\kappa L]}$$

where χ_ϕ is the character of the group element ϕ in that representation, i.e., the trace of the matrix assigned to ϕ, and $|...|$ is the norm of the character, a real number. For systems that include compact groups, the sum may be an appropriate generalized integral. The most direct assumption is that the representation is 'faithful', having as many matrices as there are group elements, but this may not be necessary.

The central idea is that F in the construct

$$\exp[-F/\kappa L] = \sum_{\nu} \exp[-|\chi_{\nu}|/\kappa L]$$

will be a Morse Function in L analogous to free energy to which we can apply Landau's classic arguments on phase transition. Again, the underlying idea is that, as the temperature of a physical system rises, more symmetries of the Hamiltonian become accessible, and this often takes place in a punctuated manner. Again, as the temperature declines, these changes are characterized as 'spontaneous symmetry breaking'. Here, we take the IDP frond length L as a temperature index, and postulate punctuated changes in IDP function and reaction dynamics with its magnitude.

Given the powerful generalities of Morse Theory, however, virtually any good Morse Function will produce spontaneous symmetry breaking under these circumstances, i.e., the behavior of interest is not restricted to a Morse Function based on the system Hamiltonian.

Following Kahraman (2009), the observed 'sloppiness' of biological lock/key molecular reaction dynamics suggests that binding site symmetry may be greater than binding ligand symmetries: binding ligands may be expected to involve (dual, mirror) *subgroups* of the nonrigid group symmetries of the IDP frond. Thus the symmetry breaking/making argument becomes

$\uparrow L \rightarrow$ more flexibility $\rightarrow$ larger binding site nonrigid symmetry group $\rightarrow$ more subgroups of possible binding sites for ligand attachment.

'Fuzzy lock theory' emerges by supposing the 'duality' between a subgroup of the IDP and its binding site can be expressed as

$$\mathcal{B}_{\alpha} = C_{\beta}\mathcal{D}_{\gamma}$$

where $\mathcal{B}_{\alpha}$ is a subgroup (or set of subgroups) of the IDP nonrigid symmetry group, $\mathcal{D}_{\gamma}$ a similar structure of the target molecule, and C_{β} is an appropriate inversion operation *or set of them* that represents static or dynamic matching of the fuzzy 'key' to the fuzzy 'lock', in the sense of Tompa and Fuxreiter (2008).

Following the taxonomy of their Table 1, if C is a single element, and $\mathcal{B}, \mathcal{D}$ fixed subgroups, then the matching would be classified as 'static'.

Increasing the number of possible elements in C, or permitting larger sets representing $\mathcal{B}$ and $\mathcal{D}$, leads to progressively more 'random' structures in an increasingly dynamic configuration, as the system shifts within an ensemble of possible states, or, perhaps, even a quantum superposition of them.

A complete treatment probably requires a groupoid generalization of nonrigid molecule theory – extension to 'partial' symmetries like those of elaborate mosaic tilings, particularly for the target species. This approach has been highly successful in stereochemisty, but remains to be done for nonrigid molecule theory.

The essential point, from the perspective of this monograph, is that the dynamic functioning of intrinsically disordered proteins can be described in terms of very elaborate symmetries indeed – those of nonrigid molecules built from wreath products of simpler groups – or their even more complex groupoid generalizations. Amyloid condensation to the steric zipper represents the collapse of those elaborate *symmetries-of-function* into much simpler conformations.

Glycan/lectin interaction

Glycan/lectin interaction at the surface of the cell, however, follows a different characteristic pattern. The glycan/lectin logic switch – the so-called 'glycosynapse' – has a markedly different dynamic from the IDP/IDR logic gate: no fuzzy-lock-and-key. Nonetheless, metabolic free energy considerations still apply.

An example. The carbohydrate α-GalNAc interacts with the lectin biotinylated soybean agglutinin (SBA) in solution to form a sequence of increasingly complicated interlinked conformations at appropriate concentrations of reacting species. Dam et al. (2007) describe this 'bind-and-slide' process in terms of a change in topology, according to figure 2.2.

Initially, the lectin diffuses along (and off) the glycan kelp frond until a number of sites are occupied. Then the lectin-coated glycan fronds begin to cross bind, until the reaction saturates in a kind of inverse spontaneous symmetry breaking. Figure 2.2D shows an end-on view of the complex shown longitudinally in figure 2.2C.

Dam and Brewer (2008) generalize:

> The bind-and-slide model for lectins binding to multivalent glycosides, globular, and linear glycoproteins is distinct from the classical 'lock and key' model for ligand-receptor interactions. The bind and slide (internal diffusion) model allows a small fraction of bound lectin

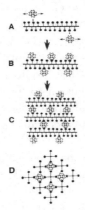

Fig. 2.2 From Dam et al. (2007). (A) At first, lectin diffuses along and off the glycan kelp frond, until, (B), a sufficient number of sites are occupied. Then (C), the lectin-coated glycan fronds begin to cross bind and the reaction is saturated, and the gate thrown. (D) shows an end-on view of the complex in (C). A kind of spontaneous symmetry breaking with increasing lectin concentration is evident, with mode (A) far 'freer' than the locked-in state of modes (C) and (D). Details will vary with the particular glycan kelp frond and the impinging lectin species.

molecules to dynamically move from carbohydrate to carbohydrate epitope in globular and linear glycoproteins. This, in turn, can facilitate lectin-mediated cross-linking of such glycoproteins on the surface of cells... Such cross-linked receptors, in turn, trigger signal transduction mechanisms... Indeed, a large number of transmembrane receptors are found clustered... Thus the affinity and hence specificity of ligand-receptor interactions may be regulated by epitope and receptor clustering in many biological systems.

Under typical physiological circumstances, glycans form a literal kelp bed bound to cellular surfaces, and the essential topological 'intensity parameter' – the temperature analog – becomes area density of the fronds. See Dam and Brewer (2010) for details. Oyelaran et al. (2009), for example, conducted density-dependent fluorescence experiments, and it was possible to take the observed intensity of that fluorescence as an index of chemical information channel capacity and switch operation, since no information transmission indicates no reaction, producing no fluorescence.

Note that, from the perspective of Oyelaran et al., figure 2.2 could be reinterpreted as displaying a spontaneous symmetry breaking with increasing 'kelp frond' area concentration at a given lectin concentration: from the relatively free modes of 2.2A, to the suddenly locked-in 'on' state of 2.2C and 2.2D. That is to say, the 'code' implied by figure 2.2 is necessarily different from the 'codes' implied by the amyloid condensation or the IDP logic gate. Nevertheless, something much like equation (2.10) will dominate the stability of the basic switching mechanisms at the cell surface, since the transmission of information is, often critically, a matter of free energy availability and control signal strength.

While IDP/IDR switches appear directly amenable to a direct symmetry analysis, the understanding of glycosynapses at the cell surface requires a deeper analysis, both because of the apparently cognitive processes producing the glycan kelp bed itself, and the different mechanism of the glycan/lectin switch itself (Wallace 2012c).

2.7   The groupoid structure of cognition

The previous sections examined protein folding and its dysfunctions largely from the perspective of relatively simple chemical dynamics, characterized by explicit symmetries. As the glycosynapse implies, however, matters rapidly escape even the complexity of group wreath products. In reality, even amyloid and prion diseases are complex *in vivo* phenomena. As Greenwald and Reik (2010) put it,

> Considering the often delicate balance between properly folded and misfolded proteins as well as the widespread occurrence of toxic protein aggregates in biology, it is not surprising that complex regulatory systems have evolved to maintain the balance of functional proteins necessary for life. These systems, collectively referred to as protein homeostasis, or 'proteostasis'... comprise a vast network that ranges from synthesis and folding (i.e., ribosome, chaperones, aggregases, disaggregases) to degradation (i.e., porteases, autophagy, lysosomal targeting). Outside of this proteostasis network, functional aggregates have their own controls that can involve external regulatory elements (i.e., enzymatic activation) or self-contained, autoregulatory elements.

We will show that the earlier approach of explicit symmetries can be extended to the larger-scale machinery that uses deterministic-but-for-error biological or other codes – or IDP components – as relatively simple elements in more complex essential control systems. That is, as many have argued, the living state involves cognitive processes at every scale of organization (Maturana and Varela 1980; Wallace 2012a, 2014a). Many forms of cognition are associated with groupoid-characterized dual information sources, significantly extending the symmetry arguments above. The development is surprisingly direct.

Atlan and Cohen (1998) argue that the essence of cognition involves comparison of a perceived signal with an internal, learned or inherited picture of the world, and then choice of one response from a much larger repertoire of possible responses. That is, cognitive pattern recognition-and-response proceeds by an algorithmic combination of an incoming external sensory signal with an internal ongoing activity – incorporating the internalized picture of the world – and triggering an appropriate action based on a decision that the pattern of sensory activity requires a response.

Incoming 'sensory' input is thus mixed in an unspecified but systematic manner with internal 'ongoing' activity to create a path of combined signals $x = (a_0, a_1, ..., a_n, ...)$. Each a_k thus represents some functional composition of the internal and the external. An application of this perspective to a standard neural network is given in Wallace (2005a).

This path is fed into some unspecified 'decision function', h, generating an output $h(x)$ that is an element of one of two disjoint sets B_0 and B_1 of possible system responses. Let

$$B_0 \equiv \{b_0, ..., b_k\},$$

$$B_1 \equiv \{b_{k+1}, ..., b_m\}.$$

Assume a graded response, supposing that if

$$h(x) \in B_0,$$

the pattern is not recognized, and if

$$h(x) \in B_1,$$

the pattern is recognized, and some action $b_j, k + 1 \leq j \leq m$ takes place.

Interest focuses on paths x triggering pattern recognition-and-response: given a fixed initial state a_0, examine all possible subsequent paths x beginning with a_0 and leading to the event $h(x) \in B_1$. Thus $h(a_0, ..., a_j) \in B_0$ for all $0 \leq j < m$, but $h(a_0, ..., a_m) \in B_1$.

For each positive integer n, take $N(n)$ as the number of high probability paths of length n that begin with some particular a_0 and lead to the condition $h(x) \in B_1$. Call such paths 'meaningful', assuming that $N(n)$ will be considerably less than the number of all possible paths of length n leading from a_0 to the condition $h(x) \in B_1$.

Identification of the 'alphabet' of the states a_j, B_k may depend on the proper system coarse-graining in the sense of symbolic dynamics.

Combining algorithm, the form of the function h, and the details of grammar and syntax, are all unspecified in this model. The assumption permitting inference on necessary conditions constrained by the asymptotic limit theorems of information theory is that the finite limit $H \equiv \lim_{n\to\infty} \log[N(n)]/n$ both exists and is independent of the path x. Again, $N(n)$ is the number of high probability paths of length n.

Call such a pattern recognition-and-response cognitive process *ergodic*. Not all cognitive processes are likely to be ergodic, implying that H, if it exists, may be path dependent, although extension to nearly ergodic processes, in a certain sense, seems possible (Wallace 2005a).

Invoking the Shannon-McMillan Theorem, it becomes possible to define an adiabatically, piecewise stationary, ergodic information source $\mathbf{X}$ associated with stochastic variates X_j having joint and conditional probabilities

$$P(a_0, ..., a_n), P(a_n | a_0, ..., a_{n-1})$$

such that appropriate joint and conditional Shannon uncertainties satisfy the classic relations of the Shannon-McMillan Theorem of Chapter 1, i.e.,

$$
\begin{aligned}
H[\mathbf{X}] &= \lim_{n\to\infty} \frac{\log[N(n)]}{n} \\
&= \lim_{n\to\infty} H(X_n | X_0, ..., X_{n-1}) \\
&= \lim_{n\to\infty} \frac{H(X_0, ..., X_n)}{n+1}
\end{aligned}
\tag{2.11}
$$

This information source is defined as *dual* to the underlying ergodic cognitive process.

'Adiabatic' means that, when the information source is properly parameterized, within continuous 'pieces', changes in parameter values take place slowly enough so that the information source remains as close to stationary and ergodic as needed to make the fundamental limit theorems work. 'Stationary' means that probabilities do not change in time, and 'ergodic' that cross-sectional means converge to long-time averages. Between pieces, it is necessary to invoke phase change formalism, the 'biological' renormalization of Chapter 1 that generalizes Wilson's (1971) approach to physical phase transition.

Again, Shannon uncertainties $H(...)$ are cross-sectional law-of-large-numbers sums of the form $-\sum_k P_k \log[P_k]$, where the P_k constitute a probability distribution.

We are not, however, constrained in this approach to the Atlan-Cohen model of cognition that, through the comparison with an internal picture of the world, invokes representation. The essential inference is that a broad class of cognitive phenomena – with and without representation – can be associated with a dual information source. The argument is direct, since cognition inevitably involves choice, choice reduces uncertainty, and this implies the existence of an information source.

For cognitive systems, an equivalence class algebra can be now constructed by choosing different origin points a_0, and defining the equivalence of two states a_m, a_n by the existence of high probability meaningful paths connecting them to the same origin point. Disjoint partition by equivalence class, analogous to orbit equivalence classes for a dynamical system, defines a groupoid. See the Mathematical Appendix for a summary of material on groupoids. This is a weak version of a very standard argument in algebraic topology leading to the definition of fundamental and free groups (Lee 2000; Crowell and Fox 1963). One might call this construction the fundamental groupoid of the cognitive process.

The vertices of the resulting network of cognitive dual languages interact to actually constitute the system of interest. Each vertex then represents a different information source dual to a cognitive process. This is not a representation of a network of interacting physical systems as such, in the sense of network systems biology. It is an abstract set of language-analogs dual to the set of cognitive processes of interest, that may become linked into higher order structures through crosstalk.

Characterization of cognition in terms of groupoids can, perhaps, be extended through application of the groupoid version of the Seifert-van Kampen Theorem (Brown et al. 2011). The question is how, when a number of cognitive processes both operate simultaneously and interact, does the groupoid associated with the joint information source relate to those of the underlying cognitive processes. The canonical example might be the global workspace of consciousness (Wallace 2005a), but wound healing and the immune response provide other examples (Wallace 2012a).

Each cognitive process X_j can be associated with an individual source uncertainty H_j. Then, by the information theory chain rule, $H[X_1, ...] \leq H[X_1] + ...$ as in equation (2.1), presumably leading to some groupoid version of equation (2.2). The choice of a fixed a_0 state as a starting point for all processes means that they all 'touch' at that base point, and this may permit definition of some appropriate 'free groupoid' in the spirit that a topological free group can be defined if several topological spaces touch at a basepoint. The resulting free group analog then would characterize the symmetry of the joint uncertainty in terms of the groupoids of the underlying sources. Details, however, do not appear to be at all simple (e.g., Baianu et al. 2005).

2.8 A Morse Function for biocognition

As briefly touched upon above, topology has become an object of algebraic study – algebraic topology – via the fundamental underlying symmetries of geometric spaces. Rotations, mirror transformations, simple ('affine') displacements, and the like, uniquely characterize topological spaces, and the networks inherent to cognitive phenomena having dual information sources also have complex underlying symmetries. Again, characterization via equivalence classes defines a groupoid, an extension of the idea of a symmetry group, as summarized by Brown (1987) or Weinstein (1996). Linkages across this set of languages occur via the groupoid generalization of Landau's spontaneous symmetry breaking arguments. As above, we use a standard approach to constructing a Morse Function parameterized in the rate of available metabolic free energy.

With each subgroupoid G_i of the fundamental groupoid associated with the cognitive process of interest we can associate source uncertainty $H(X_{G_i}) \equiv H_{G_i}$, where X is the dual information source of the cognitive phenomenon of interest.

Responses of a cognitive system can now be represented by high probability paths connecting 'initial' multivariate states to 'final' configurations, across a great variety of beginning and end points. This creates a similar variety of groupoid classifications and associated dual cognitive processes in which the equivalence of two states is defined by linkages to the same beginning and end states. Thus it becomes possible to construct a 'groupoid free energy' driven by the quality of available metabolic free energy, represented by the mitochondrial rate $\mathcal{H}$, to be taken as a temperature analog.

The argument-by-abduction from physical theory is that $\mathcal{H}$ constitutes a kind of thermal bath for the processes of biological cognition. Thus we can construct another Morse Function by writing a pseudo-probability for the information sources X_{G_i} having source uncertainties H_{G_i} as

$$\Pr[H_{G_i}] = \frac{\exp[-H_{G_i}/\kappa\mathcal{H})]}{\sum_j \exp[-H_{G_j}/\kappa\mathcal{H}]} \tag{2.12}$$

where κ is an appropriate constant characteristic of the particular system. The sum is over all possible subgroupiods of the largest available cognitive groupoid. Note that compound sources, formed by the (tunable, shifting) union of underlying transitive groupoids, being more complex, will have higher free-energy-density equivalents than those of the base transitive groupoids.

The Morse Function defined for invocation of Pettini's topological hypothesis or Landau's spontaneous symmetry breaking is then a 'groupoid free energy' $\mathcal{F}$ given by

$$\exp[-\mathcal{F}/\kappa\mathcal{H}] \equiv \sum_j \exp[-H_{G_j}/\kappa\mathcal{H}] \tag{2.13}$$

Spontaneous symmetry breaking arguments are invoked here in terms of the groupoid associated with the set of dual information sources.

Many other Morse Functions might be constructed, for example simply based on representations of the underlying cognitive groupoid(s). The resulting qualitative picture would not be significantly different.

The essential point is that decline in the rate of available mitochondrial free energy $\mathcal{H}$, or in the ability to actually use that free energy as indexed by κ, can lead to punctuated decline in the complexity of cognitive process within the entity of interest, according to this model.

If $\kappa\mathcal{H}$ is relatively large – a rich and varied real-time free energy environment – then there are many possible cognitive responses. If, however, constraints of mitochondrial aging limit the magnitude of $\kappa\mathcal{H}$, then an essential cognitive system may or will begin to collapse in a highly punctuated manner to a kind of ground state in which only limited responses are possible, represented by a simplified cognitive groupoid structure, recognizably akin to amyloid collapse in the much simpler deterministic-but-for-error protein coding machineries.

2.9 Distortion as order parameter

The Rate Distortion Function (RDF) is the minimum rate of information transmission necessary to ensure that the average distortion between message sent and message received, using a particular distortion measure over a given channel, is less than $D \geq 0$. Usually written $R(D)$, it is always a decreasing convex function of D, a reverse J-shaped curve (Cover and Thomas 2006). For example, a Gaussian channel under the squared distortion measure and in the presence of noise with zero mean and variance σ^2, has $R(D) = 1/2 \log[\sigma^2/D]$.

For protein folding in the cell, elaborate regulatory machinery is provided by the endoplasmic reticulum (Budrikis et al. 2014), implying the necessity of some comparison between what is desired and what is produced. In general, Maturana-like cognitive processes at every scale and level of organization of the living state must have regulatory systems that make similar comparisons. What we have argued in the previous two sections can be restated in terms of the collapse of the RDF with decreasing available metabolic free energy, or rather, via convexity, as the sudden appearance of a large average distortion D, as an analog to the usual order parameter in a physical system. That is, in the way magnetization disappears above a certain critical temperature in a ferromagnet, the average distortion declines in a punctuated manner in the presence of high enough rates of available metabolic free energy, driven by the underlying groupoid structure, remembering that the simplest groupoid is the disjoint union of groups, including a set consisting of a single group.

In a sense, this result can be viewed as a principled reconfiguration and extension of Friston's (2010) 'free energy' model which, in effect, minimizes the disjunction between the predictions of an inner model of the world and the returning sensory image of that world.

2.10 Biological automata are not static

It can be argued that the development of Section 2.7, via the decision function h, generalizes a standard approach from the formal theory of automata. To paraphrase Beaudry et al. (2005), finite semigroups – finite sets equipped with a binary associative operation – have played a central role in theoretical computer science for fifty years, first closely related to finite automata by Kleene (1956). Following work by Schutzenberger and Eilenberg, semigroups and automata are so tightly intertwined that it makes little sense to study one without the other. When the axiom of associativity is dropped, one obtains groupoids, with a more powerful analog to Kleene's theorem relating groupoids to push-down automata and context-free languages (Mezei and Wright, 1967). That is, context-free languages which are to be associated with groupoids are far richer than the 'regular' languages associated with semigroups. The basic flavor of automata theory, for the simple finite deterministic case, is given in the Mathematical Appendix.

Most essentially, each class of automata is linked with a different kind of well-defined dual 'language', in a very limited sense, which it recognizes in a complementarily precise manner. Here, we have weakened the underlying structure, invoking cognitive processes having rich dual 'languages' constrained by the asymptotic limit theorems of information theory. This can be interpreted as instantiating tuning around the no free lunch condition: Wolpert and MacReady (1995, 1997) established that computational optimizers all have the same average efficiency. As a consequence, an optimizer designed to be particularly good on one characteristic class of problems will be particularly bad on a complementary class. Similar results follow from the 'tuning theorem' variant of the Shannon coding theorem described in Chapter 1 in which the message has a fixed probability distribution and the channel probabilities are tuned so as to maximize the rate of information transmission. Shannon (1959), as described in Chapter 1, argues that the Rate Distortion Function, in a similar way, mandates a channel structure that minimizes the average distortion in a transmitted message. These asymptotic limit theorems thus permit description of 'biological' automata significantly less constrained than the systems characterized by the Kleene and by the Mezei and Wright theorems.

Figure 2.3 shows a schematic of such a system, representing the temporary recruitment of cognitive modules having dual information sources $X_1, ..., X_{10}$, into two different crosstalk-linked joint information sources constituting biological automata that address two qualitatively different problems confronting an organism. Such shifting, tunable coalitions of 'unconscious cognitive modules' are thought to represent the evolutionary exaptation of crosstalk into animal consciousness, wound healing, the immune system, the cell surface glycosynapse, and so on, since, from equation (2.1), allowing crosstalk ultimately consumes less metabolic free energy than isolating information sources (Wallace 2012a).

These considerations lead to yet another central topological argument.

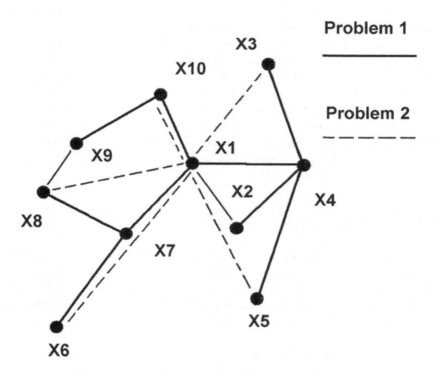

Fig. 2.3 A biological automaton is not static. Crosstalk-induced tuning of unconscious cognitive modules into different coalitions to address two different problems confronting an organism.

The crosstalk linking different cognitive modules is not inherently fixed, but is variable. This suggests a parameterized renormalization, since the modular network structure linked by that crosstalk has a topology depending on the degree of interaction of interest.

Define an interaction parameter ω, a real positive number, and look at geometric structures defined in terms of linkages set to zero if mutual information is less than, and 'renormalized' to unity if greater than, ω. Any given ω will define a regime of network elements linked by mutual information greater than or equal to it.

The argument can now be inverted. That is, a given topology for the giant component will, in turn, define some critical value, ω_C, so that network elements interacting by mutual information less than that value will be unable to participate, i.e., will be locked out and not be perceived by the system. Wallace (2005, 2012a) provides details. Thus ω is a tunable, syntactically-dependent, detection limit that depends critically on the instantaneous topology of the linked information sources defining, in effect, an analog to Baars' global broadcast of consciousness. That topology is a basic tunable syntactic filter across the underlying modular structure, and variation in ω is only one aspect of more general topological properties that can be described in terms of index theorems, where far more general analytic constraints can become closely linked to the topological structure and dynamics of underlying networks, and can stand in place of them (Hazewinkel 2002).

2.11 Finessing decision theory

The 'cognitive' argument of Sections 2.7 and 2.10 might, in some measure, be reformulated as a problem in decision theory (DT), following Dayan and Daw (2008), who examine DT methods as used across computational models in ethology, psychology, and neuroscience. They characterize two fundamental approaches, i.e., model-based, and model-free:

> Crudely speaking, model-based methods make explicit use of the actual, or learned, rules of the task to make choices. Importantly, even when the rules are fully known, it takes some computation to derive the optimal decision for a particular state from these more basic quantities.
>
> Model-free methods eschew the rules of the task and, instead, use and/or learn putatively simpler quantities that are sufficient to permit optimal choices...
>
> It has long been suggested that there is a rather direct mapping of model-free reinforcement learning algorithms onto the brain, with the neuromodulator dopamine serving as a teaching signal to train values or policies by controlling synaptic plasticity at targets such as the ventral and dorsolateral striatum...
>
> In Markov problems – that is, domains in which only the current state matters, and not the previous history – there turns out to be a computationally precise way of defining the goal for predicting future reinforcers...

A central conundrum confronting a decision theory approach to the living state, however, is that multiple decisions seem to compete simultaneously in and across organisms, acting at different scales and levels of organization. How is this to be resolved? If, following Sections 2.7 and 2.10, one assigns a dual information source to each underlying decision machine or process, then they are linked by the evolutionary exaptation of the inevitable crosstalk between them into a higher order joint information source constrained by the asymptotic limit theorems of information theory, and by the Data Rate Theorem that links information and control theories. One is not then confined to Markov systems, and the methodology subsumes underlying complications into the 'grammar' and 'syntax' of the paths associated with the dual information source(s), relying on the regularities defined by equation (2.11) and the Data Rate Theorem. One loses the detailed specificity expected from decision theory, but finesses much else.

This observation has a classic context. Recall the Heisenberg uncertainty principle: one can know either the position or momentum of a quantum system to arbitrary precision, but not both simultaneously. Similar restrictions afflict signal detection theory, involving a choice of selectivity or sensitivity, and the Bode Integral Theorem, showing that suppression of noise in one frequency range increases it in others. Here, we ignore the details of the 'grammar' and 'syntax' of the exact circumstances under which a decision is made, and look at what is left over, in a sense, and that residue is well-characterized indeed by the asymptotic limit theorems of information theory, the Data Rate Theorem, and the nonequilibrium Onsager formalism.

Parenthetically, manipulation and reinterpretation of the Bode Theorem produces the Data Rate Theorem, completing the circle, as it were (Nair et al. 2007).

Tishby and Polani (2011) reach similar conclusions:

> The information-theoretic picture is universal, general, conceptually transparent and can be post hoc imbued with the specific constraints of particular models. On the informational level, scenarios with differing computational models can be directly compared with each other. At the same time, the informational treatment allows one to incorporate limits in the information processing capacity that are fundamental properties of a particular agent-

environment system.

Tishby and Polani (2011), however, confine their calculation to Markov decision problems in which future action depends only on the current state of the system. Here, full application of the Rate Distortion and Data Rate Theorems greatly generalizes the approach.

2.12   Implications

The correspondence between unconstrained information theory inequalities and the structure of finite groups, in the context of the long-known intimate relations between semigroups, groupoids, and certain classes of formal automata, appears to foreshadow a spectrum of deeper relations between the dynamics of information sources and sometimes hidden underlying biological symmetries. These can be simple groups, as with DBFE error-minimization biological codes, or subtle 'tilings' akin to Arabic decorations – cognitive groupoids. The argument can be extended to intrinsically disordered proteins and their logic gates, via nonrigid molecular symmetries built on semidirect and wreath products of simpler groups. The satisfactory operation of such gates will then be a symmetry-constrained punctuated function of available rates (and forms) of metabolic free energy, although mathematical description of such intermediate scales is likely to be typically more difficult than the relatively simple examples described here.

Indeed, using the methods of Houghton (1975) it is possible to define wreath products of groupoids, leading to a 'nonrigid theory of cognition' – not mathematically trivial – that can be extended further via 'fuzzy' generalizations likely to better fit biological complexities (Wallace 2014b).

What seems clear is that information and symmetries, of various sorts and subtleties, may have unexpected convolutions and intertwinings, and these, in the context of the living state, will in turn be driven by the availability of metabolic free energy. Inability to provide adequate rates of, or proper forms of, that resource expresses itself in punctuated failure of central physiological function, recognizably analogous to spontaneous symmetry breaking in simple physical systems. This, via deterioration of basic cellular mitochondrial energy mechanisms, appears to be a critical component in the phenomenon of aging, and, as some have speculated may be implicated in certain classes of mental disorder, since high level cognitive function is certainly critically dependent on the proper availability of metabolic free energy (e.g., Scaglia 2010; Clay et al. 2011; Ben-Shachar

2002; Prabakaran et al. 2004; Giulivi et al. 2010; Rossignol and Frye 2014).

Attractive as the approach used here may seem, it is important to remember that these are statistical models based on the asymptotic limit theorems of information theory, much in the spirit, if not the form, of '$Y = mX + b$' regression equations.

When biologically naive physicists began plaguing ecosystem studies with oversimplified models, the mathematical ecologist E.C. Pielou (1977, pp. 107-110) replied that

> [The usefulness of mathematical models] consists *not in answering questions but in raising them.* Models can be used to inspire field investigations and these are the only source of new knowledge as opposed to new speculation... [T]he knowledge gained by model-inspired field studies leads to adjustments to the models and a further round of field studies...

The Word is not the Thing, and, like their simpler counterparts in parametric and nonparametric statistics, the models developed here will best be used as junior partners in the scientific study of similar systems under different, and different systems under similar, observational or experimental conditions.

Chapter 3

A Data Rate Theorem model

3.1 Summary

Bioregulatory failure can be formally addressed using a necessary conditions statistical model based on the Rate Distortion Theorem, and on the Data Rate Theorem that imposes a powerful interaction between information and control. Using a simplified Black-Scholes 'cost' argument, metabolic free energy, delivered at the cellular level most often in the form of ATP by healthy mitochondria, can be seen as a 'control signal' stabilizing and regulating biochemical dynamics in the presence of noise. Demand beyond an available rate of energy supply expresses itself in punctuated destabilization of the bioregulatory channel. Pathology or aging – normal, or prematurely driven by psychosocial or environmental stressors – that interferes with routine metabolic energy delivery will trigger regulatory failure leading to chronic mitochondrial disease or senescence.

3.2 Introduction

The second chapter examined Swerdlow's mitochondrial hypothesis for Alzheimer's disease (AD) (Swerdlow et al. 2010) from the perspective of symmetries associated with protein folding and biological cognition, in a large sense. The former argument involved groups associated with a deterministic error-limiting (DEL) 'protein folding code', while the latter used the groupoids associated with a certain class of cognitive processes to construct an analog to a standard Landau spontaneous symmetry breaking model of phase transition in which rates of available metabolic free energy served as a temperature index. Here, we approach the problem from the perspective of the Data Rate Theorem (Nair et al. 2007) that links information theory to control theory. The essential point is that metabolic free energy, delivered most generally in the form of ATP through the operation of healthy cellular mitochondria, must fuel any cellular control system at, and across the various levels of biological organization, and may, in the sense of Feynman (2000), actually represent the information provided by that control system. The model, involving, among other things, a biological application of Black-Scholes 'cost' methodologies, is, again, surprisingly direct. Formally, for the systems to which the asymptotic limit theorems of information and control theory apply, we show that, in the sense of Ge and Qian (2011), nonequilibrium phase transitions play a central role in living systems.

Again, we begin far afield.

Biological regulation can be viewed as the transmission of control signals to physiological systems at a variety of scales and levels of organization (Wallace, 2014b). The central questions are,

(1) how well is the control signal obeyed, and

(2) what is the intensity of metabolic free energy necessary to impose an adequate level of control.

The resulting model, from information and control theory perspectives, is certainly one of the simplest possible.

Recall the Data Rate and Rate Distortion Theorems from Chapter 1.

The Rate Distortion Theorem states that $R(D)$ is the minimum necessary rate of information transmission which ensures the communication between the biological vesicles does not exceed average distortion D. Thus $R(D)$ defines a minimum necessary channel capacity. Cover and Thomas (2006) or Dembo and Zeitouni (1998) provide details. The rate distortion function has been calculated for a number of systems, using Lagrange multiplier or Khun-Tucker optimization methods.

Cover and Thomas (2006, Lemma 13.4.1) show that $R(D)$ is necessarily a decreasing convex function of D for any reasonable definition of distortion. That is, $R(D)$ is always a reverse J-shaped curve. This will prove crucial for the overall argument. Indeed, convexity is an exceedingly powerful mathematical condition, and permits deep inference (e.g., Rockafellar 1970). Ellis (1985, Ch. VI) applies convexity theory to conventional statistical mechanics. This is the central point from which all else follows. We will use the Gaussian channel as an easily calculated example, but the arguments are quite general.

Again, for the standard Gaussian channel having noise with zero mean and variance σ^2, using the squared distortion measure,

$$R(D) = 1/2 \log[\sigma^2/D], 0 \le D \le \sigma^2$$
$$R(D) = 0, D > \sigma^2 \tag{3.1}$$

Recall the homology between information source uncertainty and free energy density. Information is a form of free energy and the construction and transmission of information within living things consumes metabolic free energy, with nearly inevitable losses via the second law of thermodynamics. If there are limits on available metabolic free energy there will necessarily be limits on the ability of living things to process information, and, in particular, for essential regulatory mechanisms to maintain the living state.

In addition, the Shannon-McMillan Theorem can be said to define a nonequilibrium steady state that can undergo nonequilibrium phase transitions in the sense of Ge and Qian (2011) analogous to, but different from those of physical systems. A similar development produces a set of Onsager-like nonequilibrium thermodynamic relations.

3.3 Black-Scholes energy cost

Suppose that metabolic free energy is available at a rate $\mathcal{H}$. Let $R(D)$ be the Rate Distortion Function describing the relation between a critical regulatory system intent and the real effect on the regulated system. This is essentially a channel capacity measure.

The distortion represents the dynamics of the disjunction between the intent of a regulatory system and its actual impact. Let R_t be the RDF of the channel connecting them at time t. The relation can, under conditions of both white noise and volatility, be expressed as the stochastic differential equation

$$dR_t = f(t, R_t)dt + bR_t dW_t \tag{3.2}$$

where dW_t represents standard white noise.

Let $\mathcal{H}(R_t, t)$ represent the rate of incoming metabolic free energy that is needed to achieve R_t at time t, and expand using the Ito chain rule (Protter, 1990):

$$d\mathcal{H}_t = [\partial\mathcal{H}/\partial t + f(R_t, t)\partial\mathcal{H}/\partial R + \frac{1}{2}b^2 R_t^2 \partial^2\mathcal{H}/\partial R^2]dt$$
$$+ [bR_t \partial\mathcal{H}/\partial R]dW_t \tag{3.3}$$

Define $\mathcal{L}$ as the Legendre transform of the free energy rate $\mathcal{H}$, a kind of entropy, having the form

$$\mathcal{L} = -\mathcal{H} + R\partial\mathcal{H}/\partial R \qquad (3.4)$$

Using the heuristic of replacing dX with ΔX in these expressions, and applying the results of equation (3.3), produces:

$$\Delta\mathcal{L} = (-\partial\mathcal{H}/\partial t - \frac{1}{2}b^2 R^2 \partial^2\mathcal{H}/\partial R^2)\Delta t \qquad (3.5)$$

Analogous to the Black-Scholes calculation (Black and Scholes, 1973), the terms in f and dW_t cancel out, so that the effects of noise are subsumed in the Ito correction involving b. Clearly, however, this invokes powerful regularity assumptions that may often be violated. Matters then revolve about model robustness in the face of such violation.

$\mathcal{L}$, as the Legendre transform of the free energy rate measure $\mathcal{H}$, is a kind of entropy that can be expected to rapidly reach an extremum at nonequilibrium steady state. There, $\Delta\mathcal{L}/\Delta t = \partial\mathcal{H}/\partial t = 0$, so that

$$\frac{1}{2}b^2 R^2 \partial^2\mathcal{H}/\partial R^2 = 0 \qquad (3.6)$$

having the solution

$$\mathcal{H}_{nss} = \kappa_1 R + \kappa_2 \qquad (3.7)$$

for appropriate constants κ_1 and κ_2, a perhaps not unexpected result.

3.4   Regulatory stability

Van den Broeck et al. (1994, 1997), Horsthemeke and Lefever (2006), and others, have noted that the relation of phase transition to driving parameters in physical systems can be obtained by using the rich stability criteria of stochastic differential equations. We examine the stability of regulation from this perspective.

The motivation follows from the observation that a Gaussian channel with noise variance σ^2 and zero mean has the Rate Distortion Function described by equation (3.1). Defining a 'Rate Distortion entropy' as the Legendre transform

$$S_R = R(D) - DdR(D)/dD = 1/2\log[\sigma^2/D] + 1/2 \qquad (3.8)$$

the simplest possible nonequilibrium Onsager equation (de Groot and Mazur 1984) is just

$$dD/dt = -\mu dS_R/dD = \mu/2D \qquad (3.9)$$

where t is the time and μ is a diffusion coefficient. By inspection, $D(t) = \sqrt{\mu t}$, the classic solution to the diffusion equation. Such 'correspondence reduction' serves as a basis to argue upward in both scale and complexity.

But regulation does not involve the diffusive drift of average distortion. Let $\mathcal{H}$ again be the rate of available metabolic free energy. Then a plausible model, in the presence of an internal system noise β^2 in addition to the environmental channel noise defined by σ^2, is the stochastic differential equation

$$dD_t = (\frac{\mu}{2D_t} - M(\mathcal{H}))dt + \frac{\beta^2}{2}D_t dW_t \qquad (3.10)$$

where dW_t represents unstructured white noise and $M(\mathcal{H}) \geq 0$ is monotonically increasing.

This has the nonequilibrium steady state expectation

$$D_{nss} = \frac{\mu}{2M(\mathcal{H})} \qquad (3.11)$$

Using the Ito chain rule on equation (3.10) (Protter 1990; Khasminskii 2012), one can calculate the variance in the distortion as $E(D_t^2) - (E(D_t))^2$. Letting $Y_t = D_t^2$ and applying the Ito relation,

$$dY_t = [2\sqrt{Y_t}(\frac{\mu}{2\sqrt{Y_t}} - M(\mathcal{H})) + \frac{\beta^4}{4}Y_t]dt + \beta^2 Y_t dW_t \qquad (3.12)$$

where $(\beta^4/4)Y_t$ is the Ito correction to the time term of the SDE.

A little algebra shows that no real number solution for the expectation of $Y_t = D_t^2$ can exist unless the discriminant of the resulting quadratic equation is ≥ 0, producing a minimum necessary rate of available metabolic free energy for regulatory stability defined by

$$M(\mathcal{H}) \geq \frac{\beta^2}{2}\sqrt{\mu} \qquad (3.13)$$

Values of $M(\mathcal{H})$ below this limit will trigger a phase transition into a less integrated – or at least behaviorally different – system in a highly punctuated manner.

From equations (3.7) and (3.11),

$$M(\mathcal{H}) = \frac{\mu}{2\sigma^2} \exp[2(\mathcal{H} - \kappa_2)/\kappa_1] \geq \frac{\beta^2}{2}\sqrt{\mu} \qquad (3.14)$$

Solving for $\mathcal{H}$ gives the necessary condition

$$\mathcal{H} \geq \frac{\kappa_1}{2} \log[\frac{\beta^2 \sigma^2}{\sqrt{\mu}}] + \kappa_2 \qquad (3.15)$$

for there to be a real second moment in D, under the subsidiary condition that $\mathcal{H} \geq 0$.

Given the context of this analysis, failure to provide adequate rates of metabolic free energy $\mathcal{H}$ would represent the onset of a regulatory stability catastrophe – in effect, a nonequilibrium phase transition. The corollary, of course, is that environmental or physiological influences increasing β, σ, the κ_i, or reducing μ, would be expected to overwhelm internal controls, triggering similar instability.

Variations of the model are possible, for example, applying the formalism to the 'natural' channel, having the rate distortion function $R(D) = \sigma^2/D$. The calculation is direct.

Equation (3.15), in fact, represents a close analog to the Data Rate Theorem of equation (1.23) (Nair et al. 2007) that relates information theory and control theory. The implication is that there is a critical rate of available metabolic free energy below which there does not exist any quantization, coding, or control scheme, able to stabilize an (inherently) unstable biological system. That is, the various versions of the DRT establish the role of nonequilibrium phase transitions in the living state (Ge and Qian 2011).

Normal, or stress-induced, aging would, at some point, be expected to affect the magnitudes of the parameters on the right hand side of the expression in equation (3.15), while simultaneously decreasing the ability to provide metabolic free energy – decreasing $\mathcal{H}$. This would result in onset of serious dysfunctions across a range of scales and levels of organization (Tomkins, 1975).

3.5   The generalized biological retina

The retina of the visual system is an example of mechanisms that may operate across a variety of physiological modes. As is well known, (e.g., Schawbe and Obermayer 2002) adaptation is a widespread phenomenon in nervous systems across multiple time scales; weeks for the activity-dependent refinement of cortical maps, hours and days for perceptual learning, seconds for the primary visual cortex. Such adaptation is a signature of an ongoing optimization of sensory systems to changing environments. Glazebrook and Wallace (2009) examine these phenomena as 'Rate Distortion manifolds', analogous to differentiable manifolds in which topologically complex local open sets are mapped one-to-one onto simple (but possibly multidimensional) plane surfaces. Here, we propose that recognizably analogous systems operate at virtually all scales and levels of organization of the living state. The essential point is to limit demand for metabolic free energy by tuning to a base state and only looking for deviations from that base. We will show, however, that even this downward projection can be energetically expensive.

Rather than taking a differential equation approach, we measure a vector of physiological responses of interest at some time $t+1$, written as X_{t+1}, that is assumed to be a function of its state at time t:

$$X_{t+1} = \mathbf{R}_{t+1}X_t \tag{3.16}$$

If X_t, the state at time t, is of dimension m, then $\mathbf{R}_t$, the manner in which that state changes in time (from time t to $t+1$), has m^2 components. If the state at time $t = 0$ is X_0, then iterating the relation above gives the state at time t as

$$X_t = \mathbf{R}_t\mathbf{R}_{t-1}\mathbf{R}_{t-2}...\mathbf{R}_1 X_0 \tag{3.17}$$

The state of the system of interest is, in this picture, essentially represented by an information-theoretic path defined by the stochastic sequence in $\mathbf{R}_t$, each member having m^2 components. That sequence is mapped onto a parallel path in the states of the physiological response, the set $X_0, X_1, ..., X_t$, each having m components.

If the state of the system can be characterized as an information source – a generalized language – so that the paths of $\mathbf{R}_t$ are autocorrelated, then the autocorrelated paths in $X(t)$ represent the output of a parallel information source that is a greatly simplified, and thus grossly distorted, picture of that system.

If the operator $\mathbf{R}$ is carefully chosen, however, his may not necessarily be the case.

Take a single iteration in more detail, assuming now that there is a (tunable) zero reference state, $\mathbf{R}_0$, for the sequence in $\mathbf{R}_t$, and that

$$X_{t+1} = (\mathbf{R}_0 + \delta\mathbf{R}_{t+1})X_t \tag{3.18}$$

where $\delta\mathbf{R}_t$ is small in some sense compared to $\mathbf{R}_0$.

Now invoke a diagonalization in terms of $\mathbf{R}_0$. Let $\mathbf{Q}$ be the matrix of eigenvectors which (Jordan) diagonalizes $\mathbf{R}_0$. Then

$$\mathbf{Q}X_{t+1} = (\mathbf{Q}\mathbf{R}_0\mathbf{Q}^{-1} + \mathbf{Q}\delta\mathbf{R}_{t+1}\mathbf{Q}^{-1})\mathbf{Q}X_t \tag{3.19}$$

If $\mathbf{Q}X_t$ is taken as an eigenvector of $\mathbf{R}_0$, say Y_k, with eigenvalue λ_k, it is possible to rewrite this equation as a spectral expansion,

$$Y_{t+1} = (\mathbf{J} + \delta\mathbf{J}_{t+1})Y_k \equiv \lambda_k Y_k + \delta Y_{t+1} =$$

$$\lambda_k Y_k + \sum_{j=1}^{n} a_j Y_j \qquad (3.20)$$

where $\mathbf{J}$ is a (block)-diagonal matrix, $\delta\mathbf{J}_{t+1} \equiv \mathbf{Q}\mathbf{R}_{t+1}\mathbf{Q}^{-1}$, and δY_{t+1} has been expanded in terms of a spectrum of the eigenvectors of $\mathbf{R}_0$, with

$$|a_j| \ll |\lambda_k|, |a_{j+1}| \ll |a_j| \qquad (3.21)$$

If $\mathbf{R}_0$ is chosen or tuned so that this condition is true, the first few terms in the spectrum of the plieotropic iteration of the eigenstate will contain most of the essential information about the perturbation. This is similar to the detection of color in the retina, where three overlapping non-orthogonal eigenmodes of response suffice to characterize a vast plethora of color sensation. Here, if such a spectral analysis is possible, a very small number of eigenmodes of system response suffice to permit identification of a vast range of perturbed physiological states: the rate-distortion constraints become very manageable indeed, and the Rate Distortion manifold simplification is an accurate characterization.

This is a complex process, since physiological systems may have both innate and learned components, and genetic programming is of limited value. The key to the problem is the proper rate-distortion tuning of the system, i.e., the choice of zero-mode, $\mathbf{R}_0$, imposed through catalysis by an embedding information source, a process that, in itself, may require considerable metabolic free energy.

That is, the arguments of equations (3.10), (3.11), and (3.13) can now be iterated, describing the dependence of the tunable base state $\mathbf{R}_0$ on available metabolic free energy. Note, however, that $M(\mathcal{H})$ may not be given by equation (3.14), since equation (3.7) is clearly only a first approximation. The essential point is that $M(\mathcal{H})$ is monotonic increasing in $\mathcal{H}$, and failure to provide adequate levels of metabolic free energy will therefore trigger failure of any biological retina.

A more formal – and much simpler – version of this argument invokes the tuning theorem form of the Shannon Coding Theorem of Section 1.1. From the perspective of the tuning theorem, equation (3.18) represents the information processing of a 'problem'. That is, X_t is 'transmitted' as a 'message' by the 'information processing channel' $\mathbf{R}_0 + \delta\mathbf{R}_{t+1}$, and recorded as the 'answer' X_{t+1}. By the tuning theorem, there will be a channel coding of $(\mathbf{R}_0 + \delta\mathbf{R}_{t+1})$ that, when properly adjusted, is most efficiently 'transmitted' by the 'problem' X_t. In general, then, the most efficient coding of the transmission channel – the best algorithm for turning a problem into a solution, is highly problem-specific. In this formulation, it is the variation of $\mathbf{R}_0$ that tunes the 'channel' to the 'problem', i.e., that transmits X_t to X_{t+1} with the least error.

Note that these mechanisms, involving focal attention through the tuning of $\mathbf{R}_0$, provide a general explanation of inattentional blindness in which overfocus on one stimulus renders an organism unable to attend to other, possibly more relevant, stimuli (e.g., Wallace 2007).

In addition, the formulation of equation (3.18) clearly falls under the constraints of the Data Rate Theorem as presented in Section 1.5, with the rate of available metabolic free energy serving as the 'control signal'.

3.6 Implications

Although equation (3.15) is built around the rate distortion function for a Gaussian channel, the argument is quite general, and some version will always emerge, driven by the convexity of the rate distortion function with the average distortion. Punctuated Data Rate Theorem instability failures, as the symmetry breaking phase transition development of Chapter 2 suggests, may be very complex indeed, most likely a characteristic sequence of 'phase change' dysfunctions in the relentless march to senescence or in the progress of chronic mitochondrial disease.

The Data Rate Theorem analog of equation (3.15) – and possible generalizations – find metabolic free energy serving as a general 'control signal' roughly in the sense of Tomkins (1975), stabilizing efficient operation of complex biochemical regulatory machinery. Demand beyond available metabolic energy supply then expresses itself in punctuated destabilization, degradation, or pathological simplification, of the regulatory channel by virtue of nonequilibrium phase transitions, a basic mechanism suggested by Ge and Qian (2011), and many others. The analysis of Chapter 2 suggests, from a different perspective, that bioregulatory phase transitions, closely associated with underlying biocognitive processes, may often involve changes in underlying groupoid symmetries, remembering that a group is a simple groupoid. In that treatment, contrasting to this, average distortion D was seen simply as an order parameter related to collapse of biological symmetries under declining rates of available metabolic free energy.

Both approaches imply that normal aging, its acceleration by psychosocial or environmental stressors, or other sources of mitochondrial dysfunction, will interfere with routine boiregulatory operation, triggering onset of many chronic diseases.

Most simply, failure of the condition of equation (3.15) can represent onset of a punctuated physiological crisis. More subtly, occurrence of a complex discriminant is often associated in physical systems with limit cycle behavior, and cycles are often characteristic of the embedding ecosystem; diurnal light/temperature variations, seasonal changes in temperature and nutrient availability, and so on. An organism might, then, avoid massive free energy demands by an evolutionary embracing, as it were, of imposed cycle behavior. Examples might include daily sleep and feeding cycles or annual patterns of mating, hibernation, and estivation, deciduous loss of leaves in plants and antlers in deer, and so on. Disruption of adapted cycling – most typically the sleep cycle in humans – then, can become a source of raised free energy demand, constituting a form of generalized inflammation more fully described in a later chapter.

Chapter 4

A mutual information model

4.1 Summary

While the symmetry-breaking and rate distortion models of the previous chapters provide, respectively, fairly comprehensive and fairly parsimonious pictures of cognitive decision dynamics, other 'greatest common factor' perspectives that act across levels of organization are also possible. Here, we explore a mutual information analysis that, while less mathematically tractable than rate distortion methods, nonetheless provides significant insight.

4.2 Formalism

Another approach to dynamic process in cognitive systems is via the mutual information generated by (the inevitable) crosstalk between channels that has been evolutionarily exapted into such processes as wound healing, the immune system, consciousness, etc. (Wallace 2012a, 2014d).

Mutual information between stochastic processes X and Y is defined as

$$I(X;Y) = H(Y) - H(Y|X)$$
$$= \sum_{x,y} p(x,y) \log[p(x,y)/p(x)p(y)$$
$$= \int_{x,y} p(x,y) \log[p(x,y)/p(x)p(y)]dxdy \qquad (4.1)$$

where the last expression is for continuous variates. It is a convex function of the conditional probability $p(y|x) = p(x,y)/p(x)$ for fixed probabilities $p(x)$ (Cover and Thomas, 2006), and this would permit a complicated construction something like that of the previous chapter, taking the x-channel as the control signal. Here, however, we treat a simplified example, involving Gaussian processes.

For two interacting information channels where the p's have normal distributions, it is tedious but easy to show that mutual information, for numerical variates, is related to correlation by the relation

$$\mathcal{M} \equiv I(X; Y) = -(1/2) \log[1 - \rho^2] \tag{4.2}$$

where ρ is the standard correlation coefficient.

Indeed, mutual information is often viewed as just another measure of correlation. This is not strictly true, however, since information is a form of free energy, particularly in biological contexts, and causal correlation between phenomena cannot occur without transfer of free energy. Thus, letting $\rho^2 \equiv Z$, we can, as in equation (3.8), define another entropy-analog as

$$S_{\mathcal{M}} \equiv \mathcal{M}(Z) - Z d\mathcal{M}(Z)/dZ \tag{4.3}$$

The simplest analog to equation (3.9) is then

$$dZ(t)/dt = \mu dS_{\mathcal{M}}/dZ = -\frac{\mu}{2} \frac{Z(t)}{(1 - Z(t))^2} \tag{4.4}$$

Carrying through the implied integral, assuming $Z(0) = 1$, gives

$$-Z^2 + 4Z - 2\log(Z) = \mu t + 3 \tag{4.5}$$

The *implicitplot* function of the computer algebra program Maple, or *ContourPlot* in Mathematica, generates figure 4.1. For small Z and large t this is becomes $Z \approx \exp[-\mu t/2]$ from the relation $dZ/dt \approx -(\mu/2)Z(t)$.

More complex empirical Onsager-like models are possible, for example

$$dZ/dt = f(dS_{\mathcal{M}}/dZ)$$
$$f = \sum_{n} -\mu_n (|dS_{\mathcal{M}}/dZ|)^n \tag{4.6}$$

where the diffusion coefficients μ_n are all positive.

The essential point is that, whatever the initial squared correlation $Z(0)$, using the gradient of $S_{\mathcal{M}}$, the nonequilibrium steady state, when $dZ/dt = 0$, is always exactly zero. That is, just like the simple diffusion solution to equation (3.9), under dynamic conditions, the final state is an uncorrelated pair of signals, in the absence of free energy exchange from crosstalk linking them.

The most direct analog to equation (3.10) is

$$dZ_t = [-\frac{\mu}{2} \frac{Z_t}{(1 - Z_t)^2} + K]dt + bZ_t dW_t \tag{4.7}$$

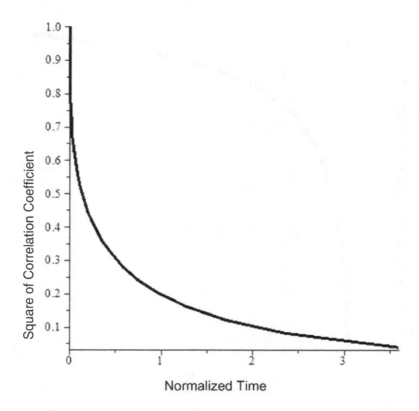

Fig. 4.1 Square of correlation coefficient between two channels characterized by normally distributed numerical variates vs. normalized time μt for mutual information without free energy crosstalk. At $t = 0$, $\rho^2 = 1$.

where K is a measure of free energy exchange between the interacting channels, μ is a diffusion rate and, again, dW_t represents white noise.

This has the steady state expectation

$$\rho^2 = \frac{1}{4} \frac{4K + \mu - \sqrt{8K\mu + \mu^2}}{K} \tag{4.8}$$

The limit of this expression, as $K \to \infty$, is exactly 1.

Figure 4.2 shows a graph of ρ^2 vs. K for $\mu = 1$. The greater the diffusion coefficient μ, the slower the rate of convergence.

It is possible to determine the standard deviation of the squared correlation (i.e., of the fraction of the joint variance) by calculating the difference

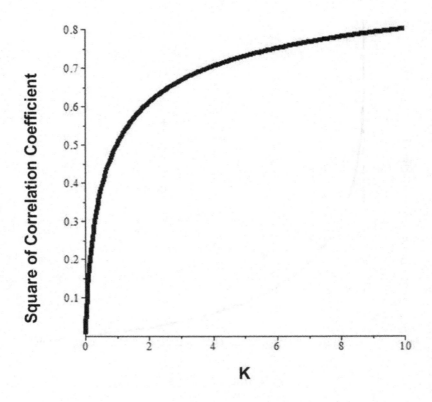

Fig. 4.2 For $\mu = 1$, the steady state expectation of the squared correlation coefficient between two linked channels having normal distributions of numerical variates. It is shown as a function of the crosstalk free energy index linking them in the model of equation (4.4). The rate of convergence of ρ^2 to 1 decreases with increasing μ.

of steady state expectations $E(Z^2) - E(Z)^2$ for the model of equation (4.7), again using the Ito chain rule.

Taking $b = \mu = 1$, there are two possible real results. One declines monotonically with increasing K, but the most evolutionarily interesting is shown in figure 4.3.

The sharp peak near $K = 0.55$, and the slow decline thereafter, suggest that relatively significant levels of crosstalk can be required to ensure stable high correlations between interacting channels having normally distributed

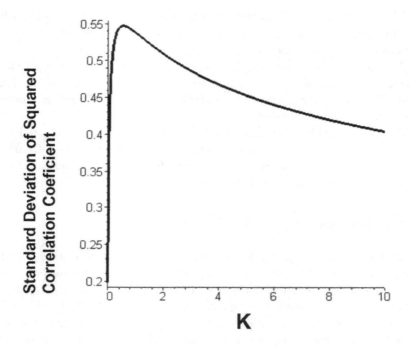

Fig. 4.3 Standard deviation at nonequilibrium steady state of the squared correlation coefficient as a function of crosstalk index K for $b = \mu = 1$ in equation (4.7). Of the real solutions, another uniformly declines with K. This, however, is the most evolutionarily interesting. Note the sharp peak near $K = 0.55$ and the relatively slow decline thereafter. Smaller values of b move the peak to the right, larger to the left. Thus, while significant levels of crosstalk are needed to ensure stable high correlation between interacting channels having normal distributions, lower correlations can be relatively stable at low free energy demand. This observation could be of importance for consciousness and analogous biological global broadcasts that are constrained by the availability of metabolic free energy. A second interpretation is that this dynamic might provide a second-order logic gate mechanism, represented by stable standard deviation in the signal to the left and right of the peak.

numerical variates. Smaller values of b move the peak to the right, higher, to the left.

This may be relevant to the degree of interaction between lower level cognitive modules needed to maintain consciousness and related 'global broadcast' biological processes (Wallace 2012a, 2014a). That is, it is not clear on which side of the peak of figure 4.3 that such systems would operate. Metabolic energy considerations suggest that it would be to the left, i.e., that global broadcasts could operate relatively stably at low crosstalk-induced channel correlations.

Another interpretation of figure 4.3 is that it might provide a physical mechanism for an instability-driven 'second order' logic gate, depending on K. The low energy state has a stable standard deviation, as does a (sufficiently) high energy condition.

For non-normally distributed numerical variates, one can expand the mutual information about a normal distribution using the Gram-Charlier method (e.g., Stuart and Ord, 1994), producing a series expression analogous to, but more complicated than, equation (4.2).

4.3   Implications

While decision theory methods are common across cognitive studies (e.g., Dayan and Daw 2008), the central problem of relating decision processes across interacting scales and levels of organization remains. Each level or scale, typically, requires a separate mathematical analysis, and linkage across them becomes difficult. Finding a greatest common factor (GCF) provides such a linkage, at the cost of an often considerable 'coarse-graining'. Information-theoretic analyses, as described in Section 2.11, are a principled method for such simplification, both in terms of groupoid topologies and rate distortion dynamics. Here, we have demonstrated that mutual information/crosstalk measures may be used to similar effect, providing a different viewpoint at the cost of some computational difficulty requiring numerical methods.

Chapter 5

A fragment size model

5.1 Summary

Earlier analysis focused on the rate of available metabolic free energy in determining the nonequilibrium steady state of a cognitive physiological process. Here, we illustrate how to extend the approach to second order rate-of-change effects in development, using renormalization methods.

5.2 Introduction

The previous three chapters have examined the direct role of available metabolic free energy rates on cognitive process, focusing on phase transitions and the stability of nonequilibrium steady states (NSS). What happens outside the NSS? Here, using the renormalization formalism of Section 1.4, we study the average fragment size, in a certain sense, to be associated with different rates of change in the rate of available metabolic free energy, using the simplest possible model.

5.3 Fragmentation dynamics

Two parameters are taken to describe the relations between an information source dual to a cognitive process and its driving metabolic free energy source.

The first, $J \geq 0$, is an inverse measure of the degree of path dependence in biological development. For systems without path dependence, $J = \infty$. $J \approx 0$ thus represents a high degree of developmental path dependence.

J will always remain distinguished, a kind of inherent direction or external field strength in the sense of Wilson (1971).

The second parameter, $Q = 1/\mathcal{H} \geq 0$, represents the inverse availability of metabolic free energy at the rate $\mathcal{H}$.

The composite structure of interest is implicitly embedded in, and operates within the context of, a larger manifold stratified by spatial, behavioral, or other 'distances'.

Take these as multidimensional vector quantities **A**, **B**, **C**.... **A** may, for example, represent location in space, **B** might represent a multivariate analysis of a spectrum of observed behavioral or other factors, in the largest sense, and so on.

It may be possible to reduce the effects of these vectors to a function of their magnitudes $a = |\mathbf{A}|$, $b = |\mathbf{B}|$ and $c = |\mathbf{C}|$, etc. Define a metric L as

$$L^2 = a^2 + b^2 + c^2 + \dots \tag{5.1}$$

Explicitly, an ergodic information source **X** represents the developmental outcome of a large-scale physiological cognitive process. The source **X**, its uncertainty $H[J, Q, \mathbf{X}]$ and its parameters J, Q depend implicitly on the embedding manifold, and in particular on the metric L.

There is a fundamental reason for adding this new layer of complication. Earlier chapters discuss the ubiquity of sudden punctuation in developmental process, i.e., relatively rapid, seemingly discontinuous fundamental changes in system structure or dynamics, but associated with the nonequilibrium steady state. The natural formalism for examination of punctuation beyond that context involves application of Wilson's (1971) program of renormalization symmetry – invariance under the renormalization transform – to source uncertainty defined on the L-manifold. The results predict that the transfer of information within a cognitive system, or between such a system and an embedding context, will undergo sudden changes in structure analogous to phase transitions in physical systems.

Much discussion of phase changes in physical phenomena is based on the assumption that, at phase transition, a system looks the same under renormalization transformation. That is, phase transition represents a stationary point for a renormalization transform in the sense that the transformed quantities are related by simple scaling laws to the original values.

Renormalization is a clustering semigroup transformation in which individual components of a system are combined according to a particular set of rules into a 'clumped' system whose behavior is a simplified average of those components. Since such clumping is a many-to-one condensation, there can be no unique inverse renormalization, and, as Section 1.4 shows, many possible forms of condensation. We will use the simplest model possible, equation (1.17). Generalization of these arguments to the models of equations (1.19) or (1.21) remains to be done.

The next step is to define characteristics of the information source **X** and J, Q as functions of averages across the manifold having metric L. That is, to 'renormalize' by clustering the entire system in terms of blocks of different sized L.

Let $N(J, Q, n)$ be the number of high probability meaningful correlated sequences of length n across the entire system in the L-manifold of equation (5.1), given parameter values J, Q. We examine changes in

$$H[J, Q, \mathbf{X}] \equiv \lim_{n \to \infty} \frac{\log[N(J, Q, n)]}{n}$$

as $Q \to Q_C$ for critical values Q_C at which the cognitive system begins to undergo a marked transformation from one kind of structure to another.

Given the metric of equation (5.1), a correlation length, $\chi(J, Q)$, can be defined as the average length in L-space over which structures involving a particular phase dominate.

Now clump the community into blocks of average size L in the multivariate manifold, the 'space' in which the cognitive system is implicitly embedded.

Following the classic argument of Wilson (1971), as in Section 1.4, it is possible to impose renormalization symmetry on the source uncertainty on H and χ by assuming at transition the relations

$$H[J_L, Q_L, \mathbf{X}] = L^D H[J, Q, \mathbf{X}] \tag{5.2}$$

and

$$\chi(J_L, Q_L) = \frac{\chi(J, Q)}{L} \tag{5.3}$$

hold, where J_L and Q_L are the transformed values of J and Q after the clumping of renormalization. Take $J_1, Q_1 \equiv J, Q$, and permit the characteristic exponent D to be nonintegral.

Equations (5.2) and (5.3) are assumed to hold in a neighborhood of the transition value Q_C.

Differentiating these with respect to L gives complicated expressions for dJ_R/dR and dQ_R/dR depending simply on L

$$dQ_L/dL = \frac{w(K_L, J_L, Q_L)}{L}$$

$$dJ_L/dL = \frac{v(K_L, J_L, Q_L)}{L} J_L \tag{5.4}$$

Solving these differential equations represents J_L and Q_L as functions of J, Q and L.

Substituting back into equations (5.2) an (5.3), and expanding in a first order Taylor series near the critical value of Q_C, gives power laws much like the Widom-Kadanoff relations for physical systems (Wilson, 1971). For example, letting $J \to 0$ and taking $\kappa \equiv (Q_C - Q)/Q_C$ gives, in first order near Q_C,

$$H = \kappa^{D/y} H_0$$
$$\chi = \kappa^{-1/y} \chi_0 \tag{5.5}$$

where y is a constant arising from the series expansion.

The 'external field strength' J remains distinguished in this treatment, i.e., the inverse of the degree of path dependence.

At the critical point, a Taylor expansion of the renormalization equations (5.2) and (5.3) gives a first order matrix of derivatives whose eigenstructure defines fundamental system behavior. For physical systems the surface is a saddle point (Wilson 1971), but more complicated behavior seems likely in cognitive process (Binney et al. 1986).

5.4　Implications

Taking the simplest formulation, $(J \to 0)$, Q increases toward a threshold value Q_C, the source uncertainty of the dual 'language' common across the cognitive system declines and, at Q_C, the average regime dominated by the 'other phase' grows. That is, the system begins to freeze into one having a large correlation length for the second phase. The two phenomena are linked at criticality in physical systems by the scaling exponent y.

Assume the rate of change of $\kappa = (Q_C - Q)/Q_C$ remains constant, $|d\kappa/dt| = 1/\tau_Q$. Analogs with physical theory suggest there is a characteristic time constant for the phase transition, $\tau \equiv \tau_0/\kappa$, such that if changes in κ take place on a timescale longer than τ for any given κ, the correlation length $\chi = \chi_0 \kappa^{-s}$, $s = 1/y$, will be in equilibrium with internal changes and result in a very large fragment in L-space.

Following Zurek (1985, 1996), the 'critical' freezout time, $\hat{t}$, will occur at a 'system time' $\hat{t} = \chi/|d\chi/dt|$ such that $\hat{t} = \tau$. Taking the derivative $d\chi/dt$, remembering that by definition $d\kappa/dt = 1/\tau_Q$, gives

$$\frac{\chi}{|d\chi/dt|} = \frac{\kappa \tau_Q}{s} = \frac{\tau_0}{\kappa} \tag{5.6}$$

so that

$$\kappa = \sqrt{s\tau_0/\tau_Q} \tag{5.7}$$

Substituting this value of κ into the equation for correlation length, the expected size of fragments in L-space, $d(\hat{t})$, becomes

$$d \approx \chi_0 (\frac{\tau_Q}{s\tau_0})^{s/2} \tag{5.8}$$

with $s = 1/y > 0$.

The more rapidly Q approaches Q_C, the smaller is τ_Q and the smaller and more numerous are the resulting Q-space fragments. Thus rapid change – sudden decline in the rate of available metabolic free energy – produces a large number of small, disjoint, cognitive fragments, where a functional system might be expected to have only a few, relatively large, individual components.

Schizophreniform disorders appear to involve developmental fragmentation of cognitive processes that should be tightly interactive. The inference is that, not only is the rate of available metabolic free energy important in development, but the second order rate of change in that rate may also have impact. For example, too rapid decline in availability of metabolic free energy during early stages of neurodevelopment may express itself in a high potential for large-scale structural fragmentation during the adolescent pruning of neural connections.

Chapter 6

Extending the perspective

6.1 Summary

The arguments leading to equations (3.8-3.10) provide a foundation for exploring more general forms of cognitive pathologies and dysfunctions that may be projected down onto basic physiological mechanisms at the cellular level. The most direct mechanism for AD emerges as a 'ground state collapse' in cognitive protein folding regulation. However, other canonical pathologies of cognitive bioregulatory process may be expressed, providing other pathways to AD, or inducing other 'developmental' pathologies such as autism and schizophrenia. Here we explore those fundamental modes, focusing on the role of 'large deviations' that may be externally or internally driven. For humans, a particularly strong external influence will be the embedding sociocultural environment and its historical trajectory.

6.2 Introduction

The 2012 Alzheimer's Association Research Roundtable (Herrup et al., 2013) observed that

> For decades, researchers have focused primarily on a pathway initiated by amyloid beta aggregation, amyloid deposition, and accumulation in the brain as the key mechanism underlying [Alzheimer's disease (AD)] and the most important target. However, evidence increasingly suggests that amyloid is deposited early during the course of the disease, even prior to the onset of clinical symptoms. Thus, targeting amyloid in patients with mild to moderate [AD],

as past failed clinical trials have done, may be insufficient to halt further disease progression. Scientists are investigating other molecular and cellular pathways and processes that contribute to AD pathogenesis...

In this chapter, we expand the perspective to include other failure modes of bioregulatory process that go beyond the previous (relatively) simple phase transition 'ground state' analysis.

6.3 Large deviations

A later section of the monumental paper by Smith et al. (2011) examines the role of large deviations in biomolecular signal transduction, and we address similar matters in our formalism.

First, the approach of equations (3.8-3.10) can be extended via an entropy-analog in a driving parameter vector $\mathbf{J}$ as the Legendre transform of a free energy-analog $\mathcal{R}$ representing some measure of information, for example a Morse Function like equation (2.13), a Rate Distortion Function, or a mutual information like equations (2.1) and (4.2),

$$S \equiv \mathcal{R}[\mathbf{J}] - \mathbf{J} \cdot \nabla_{\mathbf{J}} \mathcal{R}[\mathbf{J}] \tag{6.1}$$

Equations (3.9) and (3.10) then generalize as

$$dJ_t^k = \sum_i [L_{k,i}(t, ...\partial S/\partial J^i...)dt +$$

$$\sigma_{k,i}(t, ...\partial S/\partial J^i)dB_t^i] =$$

$$L_k(t, \mathbf{J})dt + \sum_i \sigma_{k,i}(t, \mathbf{J})dB_t^i \tag{6.2}$$

where, again, t is the time and terms have been collected and expressed in the driving parameters. The dB_t^i represent different kinds of noise – not necessarily 'white' – whose characteristics are usually expressed by their quadratic variation. See any standard text for definitions, examples, and details (e.g., Protter, 1990).

Several matters emerge directly from invoking such a coevolutionary approach:

> 1. Setting the expectation of equations (6.2) equal to zero and solving for stationary points gives attractor states since the noise terms preclude unstable equilibria.

2. This system may converge to limit cycle or pseudo-random 'strange attractor' behaviors in which the system seems to chase its tail endlessly within a limited venue – a kind of 'Red Queen' pathology.

3. What is converged to in both cases is not a simple state or limit cycle of states. Rather it is an equivalence class, or set of them, of highly dynamic information measures coupled by mutual interaction through crosstalk. Thus 'stability' in this structure represents particular patterns of ongoing dynamics rather than some identifiable static configuration. Physicists characterize this as a nonequilibrium steady state (NSS) (e.g., Derrida, 2007).

4. Applying Ito's chain rule for stochastic differential equations to $(J_t^k)^2$ and taking expectations allows calculation of variances, as in equation (3.12). These may depend very powerfully on a system's defining structural constants, leading, again, to significant instabilities defining critical phenomena whose detailed description will require the 'biological' renormalizations of Chapter 1.

This represents a highly recursive phenomenological set of stochastic differential equations (Zhu et al. 2007), but operates in a dynamic rather than static manner: information sources are inherently nonequilibrium.

Most importantly, as Champagnat et al. (2006) note, shifts between the quasi-equilibria or NSS of such a coevolutionary system can be addressed by the large deviations formalism. The dynamics of drift away from trajectories predicted by the canonical equation can be investigated by considering the asymptotic of the probability of 'rare events' for the sample paths of the diffusion.

'Rare events' are the diffusion paths drifting far away from the direct solutions of the canonical equation. The probability of such rare events is governed by a large deviation principle, driven by a 'rate function' $\mathcal{I}$ that can be expressed in terms of the parameters of the diffusion.

This result can be used to study long-time behavior of the diffusion process when there are multiple attractive singularities. Under proper conditions the most likely path followed by the diffusion when exiting a basin of attraction is the one minimizing the rate function $\mathcal{I}$ over all the appropriate trajectories.

An essential fact of large deviations theory is that the rate function $\mathcal{I}$ always has the canonical form

$$\mathcal{I} = -\sum_j P_j \log(P_j) \tag{6.3}$$

for some probability distribution. This result goes under a number of names; Sanov's Theorem, Cramer's Theorem, the Gartner-Ellis Theorem, the Shannon-McMillan Theorem, and so forth (Dembo and Zeitouni, 1998).

These arguments are in the direction of equation (6.2), now seen as subject to large deviations *that can themselves be described as the consequence of one or more information sources*, providing J^k-parameters that can trigger punctuated shifts between quasi-stable topological modes of interacting cognitive submodules. In addition, the K, κ_j, μ, β and b parameters of chapter 3 can be taken as outputs of some contextual information source.

It should be clear that both such external signals and the internal ruminations characterized by $\mathcal{I}$ can define J^k-parameters and drive the system to different nonequilibrium steady states, and do so in a highly punctuated manner. Similar ideas are now common in systems biology (e.g., Kitano 2004).

From another perspective, however, setting the expectation of equation (6.2) to zero generates an index theorem (Hazewinkel, 2002), in the sense of Atiyah and Singer (1963). Such an object relates analytic results – the solutions to the equations – to an underlying set of topological structures that are eigenmodes of a complicated network geometric operator whose spectrum is the possible states of the system. This structure, and its dynamics, do not really have simple physical system analogs.

Index theorems, in this context, instantiate relations between 'conserved' quantities – here, the quasi-equilibria of basins of attraction in parameter space – and their underlying topological forms. The argument of Chapter 2, however, described how that network was itself defined in terms of equivalence classes of meaningful paths that, in turn, defined groupoids, a significant generalization of the group symmetries more familiar to physicists.

The approach, then, in a sense – via the groupoid construction – generalizes the famous relation between group symmetries and conservation laws uncovered by E. Noether that has become the central foundation of modern physics (Byers, 1999). Thus this work proposes a kind of Noetherian statistical dynamics of the living state. The method, however, represents the fitting of dynamic regression-like statistical models based on the asymptotic limit theorems of information theory to data. These are necessary conditions models and cannot presume to be a 'real' picture of the underlying systems and their time behaviors: biology is not physics.

Again, as with simple fitted OLS regression equations, actual scientific inference is done most often by comparing the same systems under different, and different systems under the same, conditions. Statistics is not science, and one can easily imagine the necessity of nonparametric or non-Noetherian models.

6.4 Canonical pathologies of cognition

Ground state collapse

The arguments leading to equations (3.9) and (3.10) can now be recapitulated using a joint information source

$$H(X_\nu, Y, L_D) \tag{6.4}$$

providing a more complete picture of large-scale cognitive dynamics if the source X in the presence of embedding regulatory system Y, or of sporadic external 'therapeutic' or pathological interventions – or 'unexpected or unexplained internal dynamics' (UUID) – represented by the large deviations information source L_D. However, the joint source of equation (6.4) now represents a de-facto distributed cognition involving interpenetration between both the underlying embodied cognitive process and its similarly embodied regulatory machinery. We will revisit this idea in the next chapter.

That is, we can now define a composite Morse Function of embodied cognition-and-regulation, $\hat{F}$, as

$$\exp[-\hat{F}/\omega(\mathcal{H}, \mu)] \equiv \sum_i \exp[-H(X_\nu, Y, L_D)/\omega(\mathcal{H}, \mu)] \tag{6.5}$$

where $\omega(\mathcal{H}, \mu)$ is a monotonic increasing function of both the metabolic free energy (or control signal data rate) $\mathcal{H}$ and of the 'richness' of the internal cognitive function defined by an internal – strictly cognitive – network coupling parameter μ. Typical examples might include $\omega_0\sqrt{\mathcal{H}\mu}$, $\omega_0[\mathcal{H}\mu]^\gamma$, $\gamma > 0$, $\omega_1\log[\omega_2\mathcal{H}\mu + 1]$, and so on.

More generally, $H(X_\nu, Y, L_D)$ in equation (6.4) can be replaced by the norm $|\Gamma_{Y,L_D}(\nu)|$ for appropriately chosen representations Γ of the underlying cognitive-defined groupoid, in the sense of Bos (2007) and Buneci (2003). That is, many Morse Functions similarly parameterized by the monotonic functions $\omega(\mathcal{H}, \mu)$ are possible, with the underlying topology, in the sense of Pettini, itself subtly parameterized by the information sources Y and L_D.

Applying Pettini's topological hypothesis to the chosen Morse Function, reduction of either $\mathcal{H}$ or μ, or both, can trigger a 'ground state collapse' representing a phase transition to a less symmetric, pathological 'frozen' state, as in Chapter 2. In higher organisms, which must generally function under real-time constraints, elaborate secondary back-up systems have evolved to take over behavioral control under such conditions. These typically range across basic emotional, as well as hypothalamic-pituitary-adrenal (HPA) and hypothalamic-pituitary-thyroid (HPT) axis, responses (Wallace 2005a, 2012, 2014a; Wallace and Fullilove 2008). Failures of these systems are implicated across a vast range of common, and usually comorbid, mental and physical illnesses (Wallace 2005a, b; Wallace and Wallace 2010, 2013).

Again, this development represents the most direct expansion of the approach used in Chapter 2 for AD. The next iterations, however, significantly extend the general model, and may represent different possible pathways leading to a spectrum of pathologies.

'Epileptiform' disorders

An information source defining a large deviations rate function $\mathcal{I}$ in equation (6.3) can also represent input from UUID unrelated to external perturbation. Such UUID will always be possible in sufficiently large cognitive systems, since crosstalk between cognitive submodules is inevitable, and any critical value can be exceeded if the structure is large enough or is driven hard enough. This suggests that, as Nunney (1999) describes for cancer, large-scale cognitive systems must be embedded in powerful regulatory structures over the life course. Wallace (2005b), in fact, examines a 'cancer model' of regulatory failure for mental dysfunction.

Wallace (2000) uses a large deviations argument to examine epileptiform disorders. Martinerie et al. (1998) and Elger and Lehnertz (1998) find a simplified 'grammar' and 'syntax' characterize brain dynamic pathways to epileptic seizure. As Martinerie et al. put it,

> The view of chronic focal epilepsy now is that abnormally discharging neurons act as pacemakers to recruit and entrain other normal neurons by loss of inhibition and synchronization into a critical mass. Thus preictal changes should be detectable during the stages of recruitment... Nonlinear indicators may undergo consistent changes around seizure onset... We demonstrated

> that in most cases... seizure onset could be anticipated well
> in advance [using nonlinear analytic methods] and that all
> subjects seemed to share a similar 'route' towards seizure.

Wallace (2000) looks at such a phase transition in the dual source uncertainty of a spatial array of resonators itself as a large fluctuational event, having a pattern of optimal/meaningful paths defined by a (pathological) information source L_D. Failure of regulation then permits entrainment of normal neurons by abnormally discharging pacemakers, producing a seizure.

The essential point is that similar epileptiform UUID instabilities might well afflict a plethora of biological or other cognitive systems, including 'low level' physiological processes.

Obsessive/compulsive disorder

Following Overduin and Furnham (2012), obsessive-compulsive disorder (OCD) is a widespread pathological condition with prevalence rates from about 1% (current) and 2-2.5% (lifetime). Subclinical manifestations are frequently present in individuals without OCD, ranging perhaps as high as 25% of the general population. They state that

> Individuals with OCD, or with a high risk of develop-
> ing OCD, suffer from recurrent, unwanted, and intrusive
> thoughts (obsessions) and engage in repetitive ritualistic
> behaviors (compulsions), usually aimed to prevent, reduce,
> or eliminate distress or feared consequences of the obses-
> sions. Relief by rituals is generally temporary and con-
> tributes to future ritual engagement... [U]ntreated symp-
> toms often persist or increase over time, causing significant
> impairment in social, professional, academic, and/or fam-
> ily functioning...

OCD and its subclinical manifestations thus appear widespread in studied culturally Western populations, indeed, perhaps a canonical failure mode of high order cognition (but see Heine, 2001 for another possible interpretation). There may, however, be more general bioregulatory analogs at different levels of organization.

Adapting the Onsager method above to the Morse Function $\hat{F}$ of equation (6.5) leads to a generalized form of equation (6.2), now involving gradients in the extended entropy analog

$$\mathcal{S} = \hat{F}(\mathbf{Q}) - \mathbf{Q} \cdot \nabla_{\mathbf{Q}} \hat{F} \tag{6.6}$$

Again, setting the expectation the resulting set of stochastic differential equations to zero and solving for stationary sets may provide individual nonequilibrium steady states, 'Red Queen' limit cycles – where the system seems to chase its tail in repetitive cycles – or even characteristic 'strange attractor' sets over which the system engages in pseudorandom excursions.

Thus Red Queen behaviors, analogous to computer thrashing, provide a possible model of OCD in sophisticated cognitive structures that are generally both embodied and distributed. Indeed, Clay et al. (2011) invoke a mitochondrial mechanism in the cycling of bipolar disorder. Some forms of this failure mode likely act at the cellular level of organization.

Another possible route to OCD might involve a pathological 'inattentional blindness' in the spirit of Section 3.5. In that case, a complex 'cognitive retina' becomes unable to refocus on changing patterns of challenge and opportunity, and becomes fixated on a particular behavior set.

Both mechanisms may act independently or, more likely, in a complex synergistic manner.

Schizophreniform and autism spectrum analogs

Recall that the global workspace model of consciousness (Baars 1988; Baars et al. 2013; Wallace 2005a, b, 2007, 2012) posits a theater spotlight involving the recruitment of unconscious cognitive modules of the brain into a temporary, tunable, general broadcast fueled by crosstalk that allows formation of the shifting coalitions needed to address real-time problems facing a higher organism. Similar exaptations of crosstalk between cognitive modules at smaller scales have been recognized in wound healing, the immune system, and so on (Wallace 2012; 2014a).

Theories of embodied cognition envision that phenomenon as analogous, that is, as the temporary assembly of interacting modules from brain, body, and environment to address real-time problems facing an organism. This is likewise a dynamic process that sees many available information sources – not limited to those dual to cognitive brain or internal physiological modules – again linked by crosstalk into a tunable real-time phenomenon that is, in effect, a generalized consciousness. Matters related to embodied cognition will be examined more fully in the next chapter.

The perspective has particular implications for disorders of brain connectivity like schizophrenia and autism, but represents a far more general mechanism that can project downward in scale. Figure 2.3 can now be interpreted as a schematic of a generalized physiological consciousness involving dynamic patterns of crosstalk between information sources – the X_j – representing a cognitive physiological process, its biological implementation, and an embedding environment, in no given order, and treated as fundamentally equivalent. The full and dotted lines represent recruitment of these dispersed resources (involving crosstalk at or above some fixed but tunable value) in two different topological patterns to address two different kinds of problems in real time. That is, autism spectrum and schizophreniform pathologies are widely viewed as involving failures in linkage that affect, in one way or another, the recruitment of unconscious cognitive brain modules (e.g., Wallace 2005b).

With regard to schizophrenia, Ben-Shachar (2002) writes

> [M]itochondrial impairment could provide an explanation for the tremendous heterogeniety of clinical and pathological manifestations in schizophrenia...[S]everal independent lines of evidence... suggest an involvement of mitochondrial dysfunction in [the disorder]... altered cerebral energy metabolism, mitochondrial polyplasia, dysfunction of the oxidative phosphorylation system and altered mitochondrial related gene expression. [T]he interaction between dopamine, a predominant etiological factor in schizophrenia, and mitochondrial respiration is considered as a possible mechanism underlying the hyper- and hypo-activity cycling in schizophrenia.

Prabakaran et al. (2004) claim

> The etiology and pathophysiology of schizophrenia remain unknown... Almost half the altered proteins identified [by a brain tissue protenomics analysis] were associated with mitochondrial function and oxidative stress responses... We propose that oxidative stress and the ensuing cellular adaptations are linked to the schizophrenia disease process...

Shao et al. (2008) argue that a growing body of evidence suggests that there is mitochondrial dysfunction in schizophrenia, bipolar disorder, and major depressive disorder.

Scaglia (2010) finds several lines of evidence suggesting involvement of mitochondrial dysfunction in schizophrenia, including alterations in brain energy metabolism, electron transport chain activity, and expression of genes involved in mitochondrial function, and argues that mechanisms of dysfunctional cellular energy metabolism underlie the pathophysiology of major subsets of psychiatric disorders.

Clay et al. (2011) find

> Several pieces of evidence point to an underlying dysfunction of mitochondria [in bipolar disorder and schizophrenia]: (i) decreased mitochondrial respiration; (ii) changes in mitochondrial morphology; (iii) increases in mitochondrial DNA (mtDNA) polymorphisms and in levels of mtDNA mutations; (iv) downregulation of nuclear mRNA molecules and proteins involved in mitochondrial respiration; (v) decreased high-energy phosphates and decreased pH in the brain; and (vi) psychotic and affective symptoms, and cognitive decline in mitochondrial disorders... Understanding the role of mitochondria, both developmentally as well as in the ailing brain, is of critical importance to elucidate pathophysiological mechanisms in psychiatric disorders.

Similarly, there is considerable and growing evidence for mitochondrial mechanisms in autism spectrum disorders:

Palmieri and Persico (2010) write

> Autism Spectrum Disorders are often associated with clinical, biochemical, or neuropathological evidence of altered mitochondrial function... [T]he majority of autistic patients displays functional abnormalities in mitochondrial metabolism seemingly secondary to pathophysiological triggers... [That is,] mitochondrial function may play a critical role not just in rarely causing the disease, but also in frequently determining to what extent different prenatal triggers will derange neurodevelopment and yield abnormal postnatal behavior...

Likewise, Giulivi et al. (2010) assert that

> Impaired mitochondrial function may influence processes highly dependent on energy, such as neurodevelopment, and contribute to autism... [In our study,] children with autism were more likely to have mitochondrial dysfunction, mtDNA overreplication, and mtDNA deletions

than typically developing children.

Summarizing a long series of studies, Rossignol and Frye (2014) find

> [E]vidence is accumulating that [Autism Spectrum Disorder] is characterized by certain physiological abnormalities, including oxidative stress, mitochondrial dysfunction and immune dysregulation/inflammation... [R]ecent studies have... reported these abnormalities in brain tissue derived from individuals diagnosed with ASD as compared to brain tissue derived from control individuals... suggesting that ASD has a clear biological basis with features of known medical disorders.

Goh et al. (2014), in a seminal finding, argue

> Impaired mitochondrial function impacts many biological processes that depend heavily on energy and metabolism and can lead to a wide range of neurodevelopmental disorders, including autism spectrum disorder... Although evidence that mitochondrial dysfunction is a biological subtype of ASD has grown in recent years, no study to our knowledge, has demonstrated evidence of mitochondrial dysfunction in brain tissue in vivo in a large, well-defined sample of individuals with ASD... Our use of... sensitive imaging technologies has allowed us to identify in vivo a biological subtype of ASD with mitochondrial dysfunction... [L]actate-positive voxels in our sample were detected most frequently in the cingulate gyrus, a structure that supports higher-order control of thought, emotion, and behavior, and one in which both anatomical and functional disturbances have been reported previously in ASD.

The arguments of Chapter 5 suggest that rates of change in the rate of available metabolic free energy may also play a critical role in determining the ultimate overall connectivity of physiological cognitive phenomena. In particular, rapid rate-of-decay in availability of MFE may trigger developmental cognitive fragmentation.

'Mental disorders', in a large sense, emerge, then, as a synergistic dysfunction of internal process and regulatory milieu, which above was simply characterized by the interaction between the internal and external driving parameters.

Indeed, Verduzco-Flores et al. (2012), using a detailed neural network model, observed that

> [Certain network] changes result in a set of dynamics which may be associated with cognitive symptoms associated with different neuropathologies, particularly epilepsy, schizophrenia, and obsessive compulsive disorders... [S]ymptoms in these disorders may arise from similar or the same general mechanisms...[and] these pathological dynamics may form a set of overlapping states within the normal network function..[related] to observed associations between different pathologies.

Apparently, these are robust observations across a variety of cognitive systems.

Other dysfunctions likely involve characteristic irregularities in topological connections. Thus analogous disorders might arise from similar topological failures affecting the real-time recruitment of physiological, regulatory, and environmental information sources. The central environmental role of culture in human biology means, of course, that, for humans, all such disorders are inherently culture bound syndromes, much in the spirit of Kleinman and Cohen (1997) and Heine (2001). Lower level physiological systems would seem susceptible to similar culturally-specific topological pathology.

6.5 Implications

Chapters 2 and 3 examined a particular form of regulatory failure driven by the inability of mitochondrial resources to deliver critical levels of metabolic free energy. In the context of this chapter, such failure precipitates a ground state collapse to a pathologically simple physiological dynamic. Other patterns of disease, involving more subtle dysfunctions, appear to follow from a more complete exploration of canonical failure modes of cognitive bioregulatory processes. These may provide other developmental pathways to disease, beyond, (or even synergistic with) the ground state collapse leading to amyloid formation.

The role of large deviations is central to the extension of the models.

These may arise from unexpected internal dynamics, or through perturbations from an embedding, highly-structured, environmental context. For humans, an essential context is the historically-patterned sociocultural milieu in which we all participate throughout the life course. The next chapters examine how the influence of such context can percolate down into cellular physiology, the 'basic biology' of the biomedical technician or engineer. From a larger perspective, however, one is reminded of the comment by the evolutionary anthropologist Robert Boyd that "culture is as much a part of human biology as the enamel on our teeth".

Chapter 7

Embodiment and environment

7.1 Summary

The Data Rate Theorem that establishes a formal linkage between real-time linear control theory and information theory carries deep implications for the study of embodied cognition and its dysfunctions. The stabilization of such cognition is a dynamic process deeply intertwined with it, constituting, in a sense, the riverbanks directing the flow of a stream of generalized consciousness.

7.2 Introduction

The previous chapter, via equation (6.4), invoked an interpenetration between an embodied cognitive process and its similarly embodied regulatory milieu. Here, we consider such embodiment in more detail, and explore its present intellectual context.

Varela, Thompson and Rosch (1991), in their pioneering study *The Embodied Mind: Cognitive Science and Human Experience*, asserted that the world is portrayed and determined by mutual interaction between the physiology of an organism, its sensimotor circuitry, and the environment. The essential point, in their view, being the inherent structural coupling of brain-body-world. Lively debate has followed and continues (e.g., Clark, 1998; M. Wilson, 2002; A. Wilson and S. Golonka, 2013). As Wilson and Golonka put it,

> The most exciting hypothesis in cognitive science right now is the theory that cognition is embodied... [T]he implications of embodiment are... radical... If cognition can

> span the brain, body, and environment, then the 'states
> of mind' of disembodied cognitive science won't exist to
> be modified. Cognition will instead be an extended sys-
> tem assembled from a broad variety of resources... [that]
> can span brain, body, and environment... Our behavior
> emerges from a pool of potential task resources that in-
> clude the body, the environment and... the brain...

Barrett and Henzi (2005) summarize matters as follows:

> ...[C]ognition is 'situated' and 'distributed'. Cognition
> is not limited by the 'skin and skull' of the individual... but
> uses resources and materials in the environment... The dy-
> namic social interactions of primates... can be investigated
> as cognitive processes in themselves... A distributed ap-
> proach... considers all cognitive processes to emerge from
> the interactions between individuals, and between individ-
> uals and the world.

This picture, of the recruitment of markedly disparate resources into shifting, temporary coalitions in real time to address challenges and opportunities, represents a significant extension of the model we have developed so far.

It is possible to study the synergism between cognition, embodiment, and environment through the Data Rate Theorem that formally relates control theory and information theory, and to include as well the needed regulation and stabilization mechanisms in a unitary construct that must interpenetrate in a similar manner.

Above, we have formally represented a generalized cognition in terms of dual information sources linked by crosstalk. Here, we extend the model to include environment and body.

7.3   Environment as an information source

Multifactorial cognitive and behavioral systems interact with, affect, and are affected by, embedding environments that 'remember' interaction by various mechanisms. It is possible to reexpress environmental dynamics in terms of a grammar and syntax that represent the output of an information source – another generalized language.

Some examples:

1. The turn-of-the seasons in a temperate climate, for many ecosystems, looks remarkably the same year after year: the ice melts, the migrating birds return, the trees bud, the grass grows, plants and animals reproduce, high summer arrives, the foliage turns, the birds leave, frost, snow, the rivers freeze, and so on.

2. Human interactions take place within fairly well defined social, cultural, and historical constraints, depending on context: birthday party behaviors are not the same as cocktail party behaviors in a particular social set, but both will be characteristic.

3. Gene expression during development is highly patterned by embedding environmental context via 'norms of reaction' (e.g., Wallace and Wallace 2010).

Suppose it possible to coarse-grain the generalized 'ecosystem' at time t, in the sense of symbolic dynamics (e.g., Beck and Schlogl 1993) according to some appropriate partition of the phase space in which each division A_j represent a particular range of numbers of each possible fundamental actor in the generalized ecosystem, along with associated larger system parameters. What is of particular interest is the set of longitudinal paths, system statements, in a sense, of the form $x(n) = A_0, A_1, ..., A_n$ defined in terms of some natural time unit of the system. Thus n corresponds to a characteristic time unit T, so that $t = T, 2T, ..., nT$.

Again, the central interest is in serial correlations along paths.

Let $N(n)$ be the number of possible paths of length n that are consistent with the underlying grammar and syntax of the appropriately coarsegrained embedding ecosystem, in a large sense. As above, the fundamental assumptions are that – for this chosen coarse-graining – $N(n)$, the number of grammatical paths, is much smaller than the total number of paths possible, and that, in the limit of (relatively) large n,

$$H = \lim_{n \to \infty} \log[N(n)]/n$$

both exists and is independent of path.

These conditions represent a parallel with parametric statistics. Systems for which the assumptions are not true will require specialized approaches.

Nonetheless, not all possible ecosystem coarse-grainings are likely to work, and different divisions, even when appropriate, might well lead to different descriptive quasi-languages for the ecosystem of interest. Thus, empirical identification of relevant coarse-grainings for which this theory will work may represent a difficult scientific problem.

Given an appropriately chosen coarse-graining, define joint and conditional probabilities for different ecosystem paths, having the form $P(A_0, A_1, ..., A_n)$, $P(A_n|A_0, ..., A_{n-1})$, such that joint and conditional Shannon uncertainties can be defined on them that satisfy equation (1.7).

Taking the definitions of Shannon uncertainties as above, and arguing backwards from the latter two parts of equation (1.7), it is possible to recover the first, and divide the set of ecosystem temporal paths into two subsets, one very small, containing the grammatically correct, and hence highly probable paths, that we will call 'meaningful', and a much larger set of vanishingly low probability.

7.4   Body dynamics and culture as information sources

Body movement is inherently constrained by evolutionary bauplan: snakes do not brachiate, humans cannot (easily) scratch their ears with their hind legs, fish do not breathe air, nor mammals water. This is so evident that one simply does not think about it. Nonetheless, teaching a human to walk and talk, a bird to fly, or a lion to hunt, in spite of evolution, are arduous enterprises that take considerable attention from parents or even larger social groupings. Given the basic bodyplan of head and four limbs, or two feet and wings, or of a limbless spine, the essential point is that not all motions are possible. Bauplan imposes limits on dynamics.

That is, if we coarsegrain motions, perhaps using some form of the standard methods for choreography transcription appropriate to the organism (or mechanism) under study, we see immediately that not all 'statements' possible using the dance symbols have the same probability. That is, there will inevitably be a grammar and syntax to observed body-based behaviors imposed by evolutionary or explicit design bauplan. Sequences of symbols, say of length n, representing observed motions can be segregated into two sets, the first, and vastly larger, consisting of meaningless sequences (like humans scratching their ears with their feet) that have vanishingly small probability as $n \to \infty$. The other set, consistent with underlying bauplan grammar and syntax, can be viewed as the output of an information source, in precisely the manner of the previous two sections, in first approximation following the relations of equation (1.7).

In precisely the same manner as evolutionary bauplan constrains possible sequences of motions into high and low probability sets, so too learned culture (and its associated patterns of social interaction) contextually constrain possible behaviors, spoken language, body postures, and many other phenotypes. That is, different cultures impose different probability structures, in a large sense, on essential matters of living and of the life course trajectory. Even sleep is widely discordant across cultural boundaries. Birth, marriage, death, social conflict, economic exchange, and so on, are all strongly patterned by culture, in the context of historical trajectory and social segmentation. Some discussion of these matters regarding mental disorders can be found in Kleinman and Good (1985), Dejarlais et al. (1995), and references therein. Boyd and Richerson (2005) provide a more comprehensive introduction.

More generally, as Durham (1991) argues, genes and culture are two distinct but interacting systems of heritage in human populations. Information of both kinds has potential or actual influence over behaviors, creating a real and unambiguous symmetry between genes and phenotypes on the one hand and culture and phenotypes on the other. Genes and culture are best represented as two parallel tracks of hereditary influence on phenotypes, acting, of course, on markedly different timescales. Human species' identity rests, in no small part, on on its unique evolved capabilities for social mediation and cultural transmission, creating, again, high and low probability sets of real-time behavioral sequences.

7.5 Interacting information sources

Recall the arguments of Chapter 2. From the no free lunch result, Shannon's Rate Distortion insight, or the 'tuning theorem', it becomes clear that different challenges facing any cognitive system, distributed collection of them, or interacting set of other information sources, that constitute an organism (or automaton), must be met by different arrangements of cooperating modules represented as information sources.

It is possible to make a very abstract picture of this phenomenon based on the network of linkages between the information sources dual to the individual 'unconscious' cognitive modules (UCM), and those of related information sources with which they interact. That is, a shifting, task-mapped, network of information sources is continually reexpressed: given distinct problem classes confronting the organism or automaton, there must be different wirings of the information sources, including those dual to the available UCM, with the network graph edges measured by the amount of information crosstalk between sets of nodes representing the different sources: figure 2.3.

Thus fully embodied systems, in the sense of Wilson and Golonka (2013), involve interaction between very general sets of information sources assembled into a 'task-specific device' in the sense of Bingham (1988) that is necessarily highly tunable. This mechanism represents a broad evolutionary generalization of the 'shifting spotlight' characterizing the global neuronal workspace model of consciousness (Wallace, 2005a).

Recall from Chapter 2 that a mutual information measure of cross-talk is not inherently fixed, but can continuously vary in magnitude. This suggests a parameterized renormalization: the modular network structure linked by crosstalk has a topology depending on the degree of interaction of interest.

As argued, that topology is the basic tunable syntactic filter across the underlying modular structure whose more general topological properties can be described in terms of index theorems, where analytic constraints can become closely linked to the topological structure and dynamics of underlying networks (Atyah and Singer, 1963; Hazewinkel, 2002).

7.6   Implications

Here, we have explored something of the dynamics of an embodied cognition, and of a necessarily synergistic embodied regulation. These, according to theory, inevitably involve a dynamic interpenetration among nested sets of actors, represented here as information sources. They may include dual sources to internal cognitive modules, body bauplan, environmental information, language, culture, and so on.

Adapting the arguments surrounding equation (6.5), two factors determine the possible range of real-time cognitive response, in the simplest version of the model. These are the magnitude of the environmental feedback signal and the inherent structural richness of the underlying cognitive groupoid. If that richness is lacking – if the possibility of internal connections is limited – then even very high levels of a canonical control signal may not be adequate to activate appropriate behavioral responses to important real-time feedback signals, generally following a version of the argument leading to equation (3.15).

Cognition and regulation must, then, be viewed as interacting gestalt processes, involving not just an atomized individual (or, taking an even more limited 'NIMH' perspective, just the brain of that individual), but the individual in a rich context that must include both the body that acts on the environment, and the environment that reacts on body and brain.

Application of the large deviations analysis of Section 6.3 suggests that cognitive function also occurs in the context, not only of a powerful environmental embedding, but of a specific regulatory milieu: there can be no cognition without regulation. The 'stream of generalized consciousness' represented by embodied cognition must be contained within regulatory riverbanks. The failure of such containment triggers a variety of pathologies at different scales and levels of organization, and that may indeed link across both.

Chapter 8

Chronic inflammation

8.1 Summary

The punctuated dynamics of pathology associated with failure in the delivery of metabolic free energy (MFE) occur in a larger context. In particular, excessive demand for MFE, and hence drain on mitochondrial machineries that might not be able to meet demand, can be viewed as a form of generalized inflammation triggering an avalanche of serious health consequences. The nested nature of biological cognition among humans, from the subcellular to the sociocultural, implies in particular that psychosocial stress can have cascading influence across the body and over the life course trajectory. Application to AD in 'right to work' (RTW) vs. non-RTW states of the USA supports these inferences.

8.2 Aging models

Chapters 2 and 3 examined the role of metabolic free energy, most often provided by mitochondrial dynamics, in normal and pathological physiology. Chapter 5 studied developmental fragmentation, and Chapter 6 extended the model to include the influence of 'large deviations' that can be driven by embedding context. Those larger considerations, we will show, suggest that chronic bioregulatory activation can be viewed as a kind of generalized inflammation leading to shortened lifespan. In this regard, we reexamine theories of aging and place the nested cognitive modules of human life in that context.

Theories of aging abound (e.g., Lorenzini et al. 2011). One of the most popular is disposable soma theory (Kirkwood 1977) which proposes an allocation of energy leading to a trade-off between increased lifespan and increased fertility. The trade-off manifests itself as a reduction in the ability to maintain somatic cells when energy is directed toward reproductive fitness.

A quasi-programmed theory has been proposed, in which aging is an unintended consequence of a continuing developmental program, resulting in a defined lifespan (Blagosklonny 2010).

The free radical theory of aging is based on the fact that oxidative processes are essential to life, yet the consequent and subsequent generation of free radical damage is inadequately controlled, leading to accumulated damage causing dysfunctions characteristic of aging (Harman 1956).

More recently, psychosocial stress and its physiological correlates have been implicated in premature aging. Epel et al. (2004) describe accelerated telomere shortening in response to life stress. They found that psychological stress, both perceived stress and the chronicity of stress, is associated with higher oxidative stress, lower telomerase activity, and shorter telomere length, which are known determinants of cell senescence and longevity. High stress women were found to have telomeres shorter on average by the equivalent of one decade in comparison to low stress women.

Geronimus et al. (2004), focusing on the USA, describe how 'racial'/ethnic differences in chronic morbidity and excess mortality are pronounced by middle age. Evidence of early health deterioration among Blacks and racial differences in health are evident at all socioeconomic levels. They invoke a 'weathering' hypothesis positing that Blacks experience early health deterioration as a consequence of the cumulative impact of repeated experience with social or economic adversity and political marginalization.

Following Lorenzini et al., a very broad characterization of the phenotypic changes of aging is that they represent a reduced capacity to maintain homeostasis, resulting in reduced functional capacity, increased vulnerability to multiple diseases, and a reduction in the ability to respond to stress, injury, or other perturbations.

With particular regard to homeostasis, the involvement of activated bioregulation in the etiology of chronic disease has long been recognized. Bosma-den Boer et al. (2012) argue that the number of people suffering from chronic conditions such as cardiovascular disease, diabetes, respiratory diseases, mental disorders, autoimmune diseases and cancers has increased dramatically over the last three decades. The increasing rates of these chronic systemic illnesses suggest that inflammation, caused by excessive and inappropriate innate immune system activity, is unable to respond appropriately to danger signals that are new in the context of evolution. This leads, in their view, to unresolved or chronic inflammatory activation in the body.

Kolb et al. (2010) specifically study diabetes, and Miller et al. (2009) depression, from this perspective.

Cohen et al. (2012) examine the role of chronic stress in hypothalamic-pituitary-adrenocortical axis (HPA) disorders from a similar viewpoint. Chronic psychological stress is associated with increased risk for depression, cardiovascular disease, diabetes, autoimmune diseases, upper respiratory infections, and poorer wound healing, via HPA axis dysregulation (Cohen et al. 2007; McEwen 1998).

Crowson et al. (2010) provide a similar analysis, finding that inflammation and immune dysregulation are strongly implicated in the premature aging of rheumatoid arthritis (RA) patients. Premature aging due to senescence of multiple systems, such as the immune, endocrine, cardiovascular, muscular, and nervous systems, represents an attractive biologic model that may, in part, they claim, explain the excess mortality observed in RA and other chronic diseases.

Immune and HPA function are examples of (broadly) cognitive physiological and other processes as described in Chapter 2 that characterize, and are linked across, all scales and levels of organization of the living state. Following the developments of the earlier chapters, we can view the chronic activation of such systems at low levels of incoming signals as a generalized inflammation, a form of self-attack, leading to a certain spectrum of diseases.

The necessarily multilevel and multiscale phenotypes of the resulting pathology suggests that therapy and prevention will be most effective when also focused across scales and levels of organization.

Holling (1992) explores the central role of 'keystone' levels and scales in ecosystem studies, that is, those at which perturbation will particularly resonate to both smaller/lower and larger/higher levels. He argues that ecosystems are controlled and organized by a small number of key plant, animal, and biotic processes that structure the landscape at different scales. Within any one ecosystem, the periodicities and architectural attributes of the critical structuring processes will establish a nested set of periodicities and spatial features that become attractors for other variables. The degree to which small, fast events influence larger, slower ones is critically dependent upon mesoscale disturbance processes.

In sum, Holling finds that the landscape is structured hierarchically by a small number of driving processes into a similarly small number of levels, each characterized by a distinct scale of 'architectural' texture and of temporal speed and variables.

More recently, this mechanism was rediscovered in public health by Glass and McAtee (2006), expressed in terms of 'risk regulators' acting at particular scales and levels of organization. As they put it, many studies illustrate the limits of well-intentioned interventions that treat individual health behaviors as separate from social context and from biological influences. They then explore extending the stream of causation to nested levels of biology and social organization.

It is clear from much social science that particular keystone levels for humans are the geographic neighborhood of work and residence, and the embedded social networks associated with them (e.g., Wallace and Wallace 2013). This suggests that multilevel, multiscale strategies that address more than one level of organization are likely to have far more therapeutic effect than magic bullet 'ceutical' interventions limited to molecular, cellular, organ, or individual levels of structure.

8.3   Cognitive modules and broadcasts

Although, as Maturana and Varela (1980, 1992) understood, cognition pervades the living state at all scales and levels of organization, their nesting in human life extends beyond the confines of the cell, across the whole individual, and beyond. We review something of this structure.

Protein folding regulation

The 'symmetry' discussions of the second chapter have a larger context, beyond AD. High rates of protein folding and aggregation diseases, in conjunction with observations of the elaborate cellular folding regulatory apparatus associated with the endoplasmic reticulum and other cellular structures that compare produced to expected protein forms (e.g., Scheuner and Kaufman 2008; Dobson 2003), presents a clear and powerful logical challenge to simple physical 'folding funnel' free energy models of protein folding, as compelling as these are *in vitro* or *in silico*. This suggests that a more biologically-based model is needed for understanding the life course trajectory of protein folding, a model analogous to Atlan and Cohen's (1998) cognitive paradigm for the immune system. That is, the intractable set of disorders related to protein aggregation and misfolding belies simple mechanistic approaches, although free energy landscape pictures (Anfinsen 1973; Dill et al. 2007) surely capture part of the process. The diseases range from prion illnesses like Creutzfeld-Jakob disease, in addition to amyloid-related dysfunctions like Alzheimer's, Huntington's and Parkinson's diseases, and type 2 diabetes. Misfolding disorders include emphysema and cystic fibrosis.

The role of epigenetic and environmental factors in type 2 diabetes has long been known (e.g., Zhang et al. 2009; Wallach and Rey 2009). Haataja et al. (2008), for example, conclude that the islet in type 2 diabetes shows much in common with neuropathology in neurodegenerative diseases where interest is now focused on protein misfolding and aggregation and the diseases are now often referred to as unfolded protein diseases.

Scheuner and Kaufman (2008) likewise examine the unfolded protein response in β cell failure and diabetes and raise fundamental questions regarding the adequacy of simple energy landscape models of protein folding. They find that, in eukaryotic cells, protein synthesis and secretion are precisely coupled with the capacity of the endoplasmic reticulum (ER) to fold, process, and traffic proteins to the cell surface. These processes are coupled through several signal transduction pathways collectively known as the unfolded protein response that functions to reduce the amount of nascent protein that enters the ER lumen, to increase the ER capacity to fold protein through transcriptional up-regulation of ER chaperones and folding catalysts, and to induce degradation of misfolded and aggregated protein.

In the spirit of the second chapter, many of these processes and mechanisms seem no less examples of chemical cognition than the immune/inflammatory responses that Atlan and Cohen (1998) describe in terms of an explicit cognitive paradigm, or that characterizes well-studied neural processes. Details were explored in Section 2.6.

IDP logic gates

Intrinsically disordered proteins (IDP) account for some 30% of all protein species, and perhaps half of all proteins contain significant sections that are intrinsically disordered. These species and sections appear to carry out far more functional moonlighting than do highly structured proteins. Application of nonrigid molecule theory to IDP interaction dynamics gives this result directly (Wallace 2011, 2012b), in that mirror image subgroup/subgroupoid tiling matching of a fuzzy molecular lock-and-key can be much richer for IDP since the number of possible symmetries can grow exponentially with molecule length, while tiling matching for three dimensional structured proteins is relatively limited. An information catalysis model implies that this mechanism can produce a large and subtle set of biological logic gates whose properties go far beyond digital AND, OR, XOR, etc., behaviors.

Glycan/lectin logic gates

Following Wallace (2012c), application of Tlusty's rate distortion index theorem argument (Tlusty 2007) to the glycome, the glycan 'kelp bed' that coats the cell surface, and through interaction with lectin proteins, carries the major share of biological information, produces a *reductio ad absurdum* of almost infinite 'code' complexity. Clearly, a complicated form of chemical cognition imposes constraints carried by external information on what would be a grotesquely large 'glycan code error network'. The machinery that manufactures glycan kelp fronds must itself be regulated by other levels of chemical cognition to ensure that what is produced matches what was scheduled for production. More generally, the information transmission involving interaction between glycan and lectin instantiates complex logic gate structures that themselves carry out higher order cognitive processes at the intercellular level (Wallace 2012c).

Gene expression

A cognitive paradigm for gene expression has emerged, a model in which contextual factors determine the behavior of what must be characterized as a reactive system, not at all a deterministic – or even simple stochastic – mechanical process (Cohen 2006; Cohen and Harel 2007; Wallace and Wallace 2008, 2009, 2010).

O'Nuallain (2008) puts gene expression directly in the realm of complex linguistic behavior, for which context imposes meaning. He claims that the analogy between gene expression and language production is useful both as a fruitful research paradigm and also, given the relative lack of success of natural language processing by computer, as a cautionary tale for molecular biology. A relatively simple model of cognitive process as an information source permits use of Dretske's (1994) insight that any cognitive phenomenon must be constrained by the limit theorems of information theory, in the same sense that sums of stochastic variables are constrained by the Central Limit Theorem. This perspective permits a new formal approach to gene expression and its dysfunctions, in particular suggesting new and powerful statistical tools for data analysis that could contribute to exploring both ontology and its pathologies. Wallace and Wallace (2009, 2010) apply the perspective to chronic disease.

This approach is consistent with the broad context of epigenetics and epigenetic epidemiology (Jablonka and Lamb 1995, 1998; Backdahl et al.

2009; Turner 2000; Jaenish and Bird 2003; Jablonka 2004).

Foley et al. (2009) argue that epimutation is estimated to be 100 times more frequent than genetic mutation and may occur randomly or in response to the environment. Periods of rapid cell division and epigenetic remodeling are likely to be most sensitive to stochastic or environmentally mediated epimutation. Disruption of epigenetic profile is a feature of most cancers and is speculated to play a role in the etiology of other complex diseases including asthma, allergy, obesity, type 2 diabetes, coronary heart disease, autism spectrum and bipolar disorders, and schizophrenia.

Scherrer and Jost (2007a, b) invoke information theory in their extension of the definition of the gene to include the local epigenetic machinery, a construct they term the 'genon'. Their central point is that coding information is not simply contained in the coded sequence, but is, in their terms, *provided by* the genon that accompanies it on the expression pathway and controls in which peptide it will end up. In their view the information that counts is not about the identity of a nucleotide or an amino acid derived from it, but about the relative frequency of the transcription and generation of a particular type of coding sequence that then contributes to the determination of the types and numbers of functional products derived from the DNA coding region under consideration.

The genon, as Scherrer and Jost describe it, is precisely a localized form of global broadcast linking cognitive regulatory modules to direct gene expression in producing the great variety of tissues, organs, and their linkages that comprise a living entity.

The proper formal tools for understanding phenomena that 'provide' information – that are information sources – are the Rate Distortion Theorem and its zero error limit, the Shannon-McMillan Theorem, and the Data Rate Theorem.

Immune system

Atlan and Cohen (1998) have proposed an information-theoretic – and implicitly global broadcast – cognitive model of immune function and process, a paradigm incorporating cognitive pattern recognition-and-response behaviors that are certainly analogous to, but much slower than, those of the later-evolved central nervous system.

From the Atlan/Cohen perspective, the meaning of an antigen can be reduced to the type of response the antigen generates. That is, the meaning of an antigen is functionally defined by the response of the immune system. The meaning of an antigen to the system is discernible in the type of immune response produced, not merely whether or not the antigen is perceived by the receptor repertoire. Because the meaning is defined by the type of response there is indeed a response repertoire and not only a receptor repertoire.

To account for immune interpretation, Cohen (1992, 2000) has reformulated the cognitive paradigm for the immune system. The immune system can respond to a given antigen in various ways, it has 'options'. Thus the particular response observed is the outcome of internal processes of weighing and integrating information about the antigen.

In contrast to Burnet's view of the immune response as a simple reflex, it is seen to exercise cognition by the interpolation of a level of information processing between the antigen stimulus and the immune response. A cognitive immune system organizes the information borne by the antigen stimulus within a given context and creates a format suitable for internal processing; the antigen and its context are transcribed internally into the chemical language of the immune system.

The cognitive paradigm suggests a language metaphor to describe immune communication by a string of chemical signals. This metaphor is apt because the human and immune languages can be seen to manifest several similarities such as syntax and abstraction. Syntax, for example, enhances both linguistic and immune meaning.

Although individual words and even letters can have their own meanings, an unconnected subject or an unconnected predicate will tend to mean less than does the sentence generated by their connection.

The immune system creates a language by linking two ontogenetically different classes of molecules in a syntactical fashion. One class of molecules are the T and B cell receptors for antigens. These molecules are not inherited, but are somatically generated in each individual. The other class of molecules responsible for internal information processing is encoded in the individual's germline.

Meaning, the chosen type of immune response, is the outcome of the concrete connection between the antigen subject and the germline predicate signals.

The transcription of the antigens into processed peptides embedded in a context of germline ancillary signals constitutes the functional language of the immune system. Despite the logic of clonal selection, the immune system does not respond to antigens as they are, but to abstractions of antigens-in-context, and does so in a dynamic manner across many tissues, sometimes generating large-scale global broadcasts of immune activation.

Tumor control

Nunney (1999) has explored cancer occurrence as a function of animal size, suggesting that in larger animals, whose lifespan grows as about the 4/10 power of their cell count, prevention of cancer in rapidly proliferating tissues becomes more difficult in proportion to size. Cancer control requires the development of additional mechanisms and systems to address tumorigenesis as body size increases – a synergistic effect of cell number and organism longevity. Nunney concludes that this pattern may represent a real barrier to the evolution of large, long-lived animals and predicts that those that do evolve have recruited additional controls over those of smaller animals to prevent cancer.

In particular, different tissues may have evolved markedly different tumor control strategies. All of these, however, are likely to be energetically expensive, permeated with different complex signaling strategies, and subject to a multiplicity of reactions to signals, including those related to psychosocial stress. Forlenza and Baum (2000) explore the effects of stress on the full spectrum of tumor control, ranging from DNA damage and control, to apoptosis, immune surveillance, and mutation rate. Wallace et al. (2003) argue that this elaborate tumor control strategy, in large animals, must be at least as cognitive as the immune system itself, one of its principal components: some comparison must be made with an internal picture of a healthy cell, and a choice made as to response, i.e., none, attempt DNA repair, trigger programmed cell death, engage in full-blown immune attack. This is, from the Atlan/Cohen perspective, the essence of cognition, and clearly involves the recruitment of a comprehensive set of cognitive subprocesses into a larger, highly tunable, dynamic structure across a variety of different tissue subtypes.

Wound healing

Following closely Mindwood et al. (2004), mammalian tissue repair is a series of overlapping events that begins immediately after wounding. Platelet aggregation forms a hemostatic plug and blood coagulation forms the provisional matrix. This dense cross-linked network of fibrin and fibronectin from blood acts to prevent excessive blood loss. Platelets release growth factors and adhesive proteins that stimulate the inflammatory response, entraining immune function, and inducing cell migration into the wound using the provisional matrix as a substrate. Wound cleaning is done by neutrophils, solubilizing debris, and monocytes that differentiate into macrophages and phagocytose debris. The macrophages release growth factors and cytokines that activate subsequent events. For cutaneous woulds, keratinocytes migrate across the area to reestablish the epithelial barrier.

Fibroblasts then enter the wound to replace the provisional matrix with granulation tissue composed of fibronectin and collagen. As endothelial cells revascularize the damaged area, fibroblasts differentiate into myofibroblasts and contract the matrix to bring the margins of the wound together. The resident cells then undergo apoptosis, leaving collagen-rich scar tissue that is slowly remodeled in the following months. Wound healing, then, provides an ancient example of a global broadcast that recruits a set of cognitive processes, in the Atlan/Cohen sense. The mechanism, which may vary across taxa, is inherently tunable, addressing the signal of 'excessive distortion' represented by a wound.

HPA axis

Reiterating the Atlan/Cohen argument, the essence of cognition is comparison of a perceived external signal with an internal picture of the world, and then, upon that comparison, the choice of a response from a much larger repertoire of possible responses. Clearly, from this perspective, the HPA axis, the flight-or-fight reflex, is cognitive. Upon recognition of a new perturbation in the surrounding environment, emotional and/or conscious cognition evaluate and choose from several possible responses: no action necessary, flight, fight, helplessness (flight or fight needed, but not possible). Upon appropriate conditioning, the HPA axis is able to accelerate the decision process, much as the immune system has a more efficient response to second pathogenic challenge once the initial infection has become encoded in immune memory. Certainly hyperreactivity as a sequela of post traumatic stress disorder (PTSD) is well known. Wallace and Wallace (2010, 2013) provide detailed models.

Bjorntorp (2001) in particular examines the role of chronic HPA axis activation in abdominal and visceral obesity, a matter of great clinical and public health importance.

Blood pressure regulation

Rau and Elbert (2001) review much of the literature on blood pressure regulation, particularly the interaction between baroreceptor activation and central nervous function. We paraphrase something of their analysis. The essential point, of course, is that unregulated blood pressure would be quickly fatal in any animal with a circulatory system, a matter as physiologically fundamental as tumor control. Much work over the years has elucidated some of the mechanisms involved: increase in arterial blood pressure stimulates the arterial baroreceptors which in turn elicit the baroreceptor reflex, causing a reduction in cardiac output and in peripheral resistance, returning pressure to its original level. The reflex, however, is not actually this simple: it may be inhibited through peripheral processes, for example under conditions of high metabolic demand. In addition, higher brain structures modulate this reflex arc, for instance when threat is detected and fight or flight responses are being prepared. Thus blood pressure control cannot be a simple reflex. It is, rather, a broad and actively cognitive modular system which compares a set of incoming signals with an internal reference configuration, and then chooses an appropriate physiological level of blood pressure from a large repertory of possible levels – a cognitive process in the Atlan/Cohen sense. The baroreceptors and the baroreceptor reflex are, from this perspective, only one set of a complex array of components making up a larger and more comprehensive cognitive blood pressure regulatory module.

Emotion

Thayer and Lane (2000) summarize the case for what can be described as a cognitive emotional process. Emotions, in their view, are an integrative index of individual adjustment to changing environmental demands, an organismal response to an environmental event that allows rapid mobilization of multiple subsystems. Emotions are the moment-to-moment output of a continuous sequence of behavior, organized around biologically important functions. These 'lawful' sequences have been termed 'behavioral systems' by Timberlake (1994).

Emotions are self-regulatory responses that allow the efficient coordination of the organism for goal-directed behavior. Specific emotions imply specific eliciting stimuli, specific action tendencies (including selective attention to relevant stimuli), and specific reinforcers. When the system works properly, it allows for flexible adaptation of the organism to changing environmental demands, so that an emotional response represents a *selection* of an appropriate response and the inhibition of other less appropriate responses from a more or less broad behavioral repertoire of possible responses. Such 'choice' leads directly to something closely analogous to the Atlan and Cohen language metaphor.

Thayer and Friedman (2002) argue, from a dynamic systems perspective, that failure of what they term 'inhibitory processes' which, among other things, direct emotional responses to environmental signals, is an important aspect of psychological and other disorder. Sensitization and inhibition, they claim, 'sculpt' the behavior of an organism to meet changing environmental demands. When these inhibitory processes are dysfunctional – choice fails – pathology appears at numerous levels of system function, from the cellular to the cognitive.

Gilbert (2001) suggests that a canonical form of such pathology is the excitation of modes that, in other circumstances, represent 'normal' evolutionary adaptations, a general matter to which we will return.

Panskepp (2003) has argued that emotion represents a primary form of consciousness, based in early-evolved brain structures, which has become convoluted with a later-developed global neuronal workspace. The convolution with individual consciousness appears to involve a large number of other cognitive biological and social submodules as well.

Consciousness

Sergeant and Dehaene (2004) characterize individual consciousness in terms of Bernard Baars' work, who has proposed that consciousness is associated with the interconnection of multiple areas processing a stimulus by a dynamic 'neuronal workspace' within which recurrent connections allow long-distance communication and auto-amplification of the activation. Neuronal network simulations, they claim, suggest the existence of a fluctuating dynamic threshold, as described in Section 2.10. If the primary activation evoked by a stimulus exceeds this threshold, reverberation takes place and stimulus information gains access, through the workspace, to a broad range of other brain areas allowing, among other processes, verbal report, voluntary manipulation, voluntary action and long-term memorization.

Below this threshold, they argue, stimulus information remains unavailable to these processes. Thus the global neuronal workspace theory predicts an all-or-nothing transition between conscious and unconscious perception. More generally, many non-linear dynamical systems with self-amplification are characterized by the presence of discontinuous transitions in internal state.

Thus Baars' global workspace model of animal consciousness sees the phenomenon as a dynamic array of unconscious cognitive modules that unite to become a global broadcast having a tunable perception threshold not unlike a theater spotlight, but whose range of attention is constrained by embedding contexts (Baars 1988, 2005; Baars et al. 2013). As Baars and Franklin (2003) put it:

> 1. The brain can be viewed as a collection of distributed specialized networks (processors).
>
> 2. Consciousness is associated with a global workspace in the brain – a fleeting memory capacity whose focal contents are widely distributed – 'broadcast' – to many un-

conscious specialized networks.

3. Conversely, a global workspace can also serve to integrate many competing and cooperating input networks.

4. Some unconscious networks, called contexts, shape conscious contents, for example unconscious parietal maps modulate visual feature cells that underlie the perception of color in the ventral stream.

5. Such contexts work together jointly to constrain conscious events.

6. Motives and emotions can be viewed as goal contexts.

7. Executive functions work as hierarchies of goal contexts.

The basic mechanism emerges directly from application of the asymptotic limit theorems of information theory, once a broad range of unconscious cognitive processes is recognized as inherently involving information sources – generalized languages, as discussed above (Wallace 2000, 2005a, 2007, 2012a). This permits mapping physiological unconscious cognitive modules onto an abstract network of interacting information sources, allowing a simplified mathematical attack that, in the presence of sufficient linkage – crosstalk – permits rapid, shifting, global broadcasts in response to sufficiently large impinging signals. The topology of that broadcast is tunable, depending on the spectrum of distortion measures and contextual 'riverbank' limits imposed on the system of interest.

Sociocultural cognition

Humans entertain a hypersociality that embeds us all in group decisions and collective cognitive behavior within a social network, itself embedded in a historically-structured, path-dependent, shared culture. As argued in Chapter 7, for humans, culture is utterly fundamental (Durham 1991; Richerson and Boyd 2006; Jablonka and Lamb 1995). Genes and culture represent two parallel lines or tracks of hereditary influence on phenotypes: the centrality of hominid evolution can be characterized as an interweaving of genetic and cultural systems. Genes came to encode for increasing hypersociality, learning, and language skills. The most successful populations displayed increasingly complex structures that better aided in buffering the local environment.

Successful human populations have a core of tool usage, sophisticated language, oral tradition, mythology, music, and decision making skills focused on relatively small family/extended family social network groupings. More complex social structures are built on the periphery of this basic object. The human species' very identity may rest on its unique evolved capacities for social mediation and cultural transmission. These are particularly expressed through the cognitive decision making of small groups facing changing patterns of threat and opportunity, processes in which we are all embedded and all participate.

The emergent cognitive behavior of organizations has long been studied under the label distributed cognition. Hollan et al. (2000) argue that unlike traditional cognitive theories, the theory of distributed cognition, much like embodied cognition, extends the reach of what is considered *cognitive* beyond the individual to encompass interactions between people and with resources and materials in the environment. Distributed cognition refers to a perspective on all of cognition, rather than a particular kind of cognition. Distributed cognition looks for cognitive processes, wherever they may occur, on the basis of the functional relationships of elements that participate together in the process. A process is not cognitive simply because it happens in a brain, nor is a process noncognitive simply because it happens in the interactions between many brains. Distributed cognition describes, in their view, a system that can dynamically configure itself to bring subsystems into coordination to accomplish various functions.

Wallace and Fullilove (2008) apply something of these approaches to institutions and other social structures. Wallace (2013, 2015) extends the analysis.

Bruce et al. (2009) review in great detail the role of sociocultural stress in the etiology of a spectrum of what might well be considered generalized autoimmune disorders.

8.4   Rates of generalized consciousness

Following Wallace (2012a), gene expression, wound healing, the immune response, tumor control, and animal consciousness all represent the evolutionary exaptation of crosstalk into processes that recruit sets of simpler cognitive modules into temporary working coalitions to address patterns of threat and opportunity confronting the organism. They do so, however, at markedly different rates. Wound healing, depending on the extent of injury, may take 18 months to complete its work. Animal consciousness typically operates with a time constant of a few hundred milliseconds. How can phenomena acting on such different rates be subsumed under the same rubric? A heuristic answer is relatively straightforward: neural tissues in humans consume metabolic free energy at ten times the rate of other tissues, and adaptation of the Arrhenius law, which predicts exponential differences in reaction rate with 'temperature', in a large sense, produces the result.

Indeed, the underlying energetics of biological reactions are remarkable. At 300 K, molecular energies represent about 2.5 KJ/mol in available free energy. By comparison, the basic biological energy reaction, the hydrolysis of adenosine triphosphate (ATP) to the diphosphate form, under proper conditions at 300 K, produces about 50 KJ/mol, equivalent to a 'reaction temperature' of 6000 K. Increasing the rate of ATP delivery to one kind of tissue an order of magnitude more than any others provides sufficient energy for extremely rapid biocognition.

In more detail, given a chemical reaction of the form $aA+bB \rightarrow pP+qQ$, the rate of change in (for example) the concentration of chemical species P (written $[P]$) is typically determined by an equation of the form

$$d[P]/dt = k(T)[A]^n[B]^m \tag{8.1}$$

where n and m are constants depending on the reaction details. The reaction rate k is expressed by the Arrhenius relation as

$$k = \alpha \exp[-E_a/RT] \tag{8.2}$$

where α is another characteristic constant, E_a is the reaction activation energy, T is the temperature and R a universal constant. $\exp[-E_a/RT]$ is, from the Boltzmann relation, the fraction of molecular interactions having energy greater than E_a. Figure 8.1 shows the form of this expression for $k = \exp[-1/M]$.

A similar information theory model leads to much the same result. As we have argued at some length, cognition can often be associated with a dual information source, whose source uncertainty – an inherent rate function – we designate as $H \leq C$, where C is the limiting channel capacity of the system. Consciousness is well-known to most often be an all-or-nothing phenomenon (Sergeant and Dehaene 2004), so that a cognitive process of signal detection must exceed some threshold before becoming entrained into the characteristic general broadcast. Following the 'renormalization' arguments of Section 2.10, let the threshold source uncertainty be H_0, representing the free energy of the incoming signal that carries meaning.

We can again write the probability for $H \geq H_0$ in a Boltzmann-like form, since information can be viewed as another form of free energy. As-

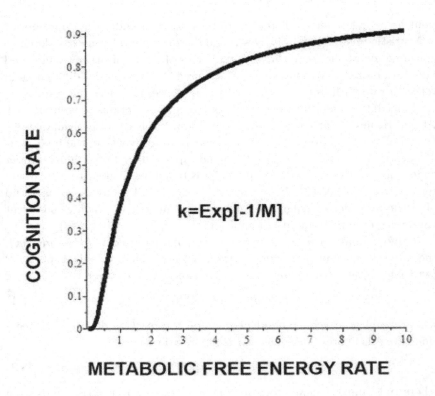

Fig. 8.1   Arrhenius-like relation for rate of a generalized consciousness or other cognitive physiological process as a function of the rate of available metabolic free energy M. Decline in M below the shoulder of the curve triggers catastrophic collapse of cognition. More complicated perspectives would treat such failure as a phase transition at a critical value of M, or else as violation of the necessary condition of the Data Rate Theorem.

suming C is quite large compared to κM gives

$$P[H > H_0] = \frac{\int_{H_0}^{C} \exp[-x/\kappa M]dx}{\int_{0}^{C} \exp[-x/\kappa M]dx} \approx \exp[-H_0/\kappa M] \qquad (8.3)$$

Taking $H_0 = \kappa = 1$ replicates figure 8.1.

Since, in mammals, body temperature remains constant, the rate of available metabolic free energy serves as the temperature analog in determining the characteristic rate of a generalized consciousness or other cognitive physiological process. Neural mechanisms, in humans having an order of magnitude greater metabolic rate than other tissues, can thus easily act at rates much greater than other physiological phenomena, although the mechanism illustrated by figure 8.1 does reach a point of diminishing returns. Another interpretation, however, is that decline in the rate of available metabolic free energy can, below the shoulder of the figure, cause sudden catastrophic collapse of cognitive function. This is, perhaps, the most parsimonious model possible for AD and other mitochondrial disorders. Other analyses treat such sudden change as a phase transition at a critical M or as violation of the stability limit of the Data Rate Theorem. See Sections 2.8 and 3.4.

8.5 Culturally-specific generalized inflammation

Extending the arguments of Wallace and Wallace (2010), it seems clear that the stresses that activate cognitive modules and their shifting coalitions are not random sequences of perturbations, and are not independent of culturally-modulated perception. Rather, stress responses involve highly correlated, grammatical, syntactical processes by which an embedding psychosocial environment communicates with an individual's cognitive physiological and mental hierarchy, strongly structured by the power relations between groups. The stress experienced by an individual can thus be taken as another adiabatically piecewise stationary ergodic (APSE) information source, interacting with a set of dual information sources defined by modular cognition at different scales and levels of organization.

Recall that the ergodic nature of the language of stress is essentially a generalization of the law of large numbers, so that long-time averages approximate cross-sectional expectations. Languages do not have simple autocorrelation patterns, in distinct contrast with the usual assumption of random perturbations by white noise in the standard formulation of stochastic differential equations.

Suppose we measure stress by determining the concentrations of HPA axis hormones and other biochemical signals according to an appropriate natural time frame, taken as the inherent period of the system. In the absence of extraordinary meaningful psychosocial stress, we measure a series of n concentrations at time t represented as an n-dimensional vector X_t. We conduct a number of experiments, and create a regression model so that, in the absence of extreme perturbation, and to first order, the concentration of biomarkers at time $t + 1$ can be written in terms of that at time t using a matrix equation of the form

$$X_{t+1} \approx < \mathbf{R} > X_t + b_0 \tag{8.4}$$

where $< \mathbf{R} >$ is the matrix of regression coefficients and b_0 a vector of constant terms.

In the presence of a modest perturbation by structured stress,

$$X_{t+1} = (< \mathbf{R} > + \delta \mathbf{R}_{t+1})X_t + b_0$$
$$\equiv < \mathbf{R} > X_t + \epsilon_{t+1} \tag{8.5}$$

where b_0 and $\delta \mathbf{R}_{t+1} X_t$ are absorbed into a vector ϵ_{t+1} of error terms *that are not necessarily small*.

It is important to realize that this is not a population process whose continuous analog is exponential growth. Rather this represents the passage of a signal – structured psychosocial stress – through a distorting physiological and/or mental filter.

If the matrix of regression coefficients $< \mathbf{R} >$ is sufficiently regular, it is possible to (Jordan block) diagonalize it using the matrix of its column eigenvectors $\mathbf{Q}$, writing

$$\mathbf{Q}X_{t+1} = (\mathbf{Q} < \mathbf{R} > \mathbf{Q}^{-1})\mathbf{Q}X_t + \mathbf{Q}\epsilon_{t+1} \tag{8.6}$$

or equivalently as

$$Y_{t+1} = < \mathbf{J} > Y_t + W_{t+1} \tag{8.7}$$

where $Y_t \equiv \mathbf{Q}X_t, W_{t+1} \equiv \mathbf{Q}\epsilon_{t+1}$, and $< \mathbf{J} > \equiv \mathbf{Q} < \mathbf{R} > \mathbf{Q}^{-1}$ is a (block) diagonal matrix in terms of the eigenvalues of $< \mathbf{R} >$.

Thus the (rate distorted) writing of structured stress on the affected physiological submodules through $\delta \mathbf{R}_{t+1}$ is reexpressed in terms of the vector W_{t+1}.

The sequence of W_{t+1} is, in analogy with the arguments of Section 3.5, the rate-distorted image of the information source defined by the system of external structured psychosocial stress. This formulation permits estimation of the nonequilibrium steady-state effects of that image on underlying cognitive physiological and mental modules and dynamics. Since everything is APSE, it becomes possible to either time or ensemble average both sides of equation (8.7), so that the one-period offset is absorbed in the averaging, giving a nonequilibrium steady state relation

$$< Y > = < \mathbf{J} > < Y > + < W > \tag{8.8}$$

or

$$< Y >= (\mathbf{I}- < \mathbf{J} >)^{-1} < W > \qquad (8.9)$$

where $\mathbf{I}$ is the $n \times n$ identity matrix.

Now reverse the argument, choosing Y_k to be a fixed eigenvector of $< \mathbf{R} >$. Expressing the diagonalization of $< \mathbf{J} >$ in terms of its eigenvalues gives the average excitation of the generalized physiological stress response in terms of an appropriate eigentransformed pattern of exciting perturbations as

$$< Y_k >= \frac{< W_k >}{1- < \lambda_k >} \qquad (8.10)$$

where $< \lambda_k >$ is the eigenvalue of $< Y_k >$, and $< W_k >$ is a similarly appropriately transformed set of ongoing perturbations by structured psychosocial stress.

In consequence, there will be a culturally characteristic form of perturbation by structured psychosocial stress – the W_k – that will resonantly excite a particular eigenmode of the generalized physiological stress response. Conversely, by tuning the eigenmodes of $< \mathbf{R} >$ – similarly to the development in Section 3.5 – the generalized stress response can be trained to galvanized excitation in the presence of particular forms of perturbation.

This is because, if $< \mathbf{R} >$ has been appropriately determined from regression relations, then the λ_k will be a kind of multiple correlation coefficient (Wallace and Wallace 2000) so that particular eigenpatterns of perturbation will have greatly amplified impact on the generalized inflammatory response. If $\lambda = 0$ then perturbation has no more effect than its own magnitude. However, if $\lambda \to 1$, then the written image of a perturbing psychosocial stressor will have very great impact. We characterize a system with $\lambda \approx 0$ as locally-resilient since its response is no greater than the perturbation itself.

Learning by the cognitive physiological modules is, it can be argued, the process of tuning response to perturbation – the generalized physiological retina of Section 3.5. This is why, here, we have written $< \mathbf{R} >$ instead of simply $\mathbf{R}$: the regression matrix is a tunable set of variables. Again, the argument could be greatly simplified by invoking the tuning theorem variant of the coding theorem.

The next section examines Alzheimer's disease in the USA from the perspective of work-related stress.

8.6   Work stress and AD

A recent study by Wang et al. (2012) suggests a mechanism for stress in the etiology of AD in Western societies. Examining a dementia-free cohort of over 900 residents aged 75+ in Stockholm over a six year period, they found that failure to have high job control doubled the risk of developing AD in later life, for this sample. They had hypothesized that psychosocial stress may contribute to the risk of dementia and AD in late life through stress-related alterations, on the basis of a glucocorticoid cascade hypothesis, and indeed found that, compared with persons who had high job control in their occupational life, persons who had low job control were at higher risk of developing dementia and AD in late life, supporting the hypothesis that high job demands and low control may increase the risk of dementia through stress rather than vascular mechanisms.

Figure 8.2 carries the argument further. Adapted from Singh-Manoux et al. (2003), it shows sex-specific dose-response curves of age-adjusted self-reported ill-health vs. an inverse self-reported status rank for the Whitehall II cohort of UK government workers, in a sense a best-case scenario. 1 is high, and 10 low, status. The curve approaches the 'LD-50' at which half the dosed population shows physiological effect.

The figure suggests, for this privileged and relatively affluent subpopulation of a Western democracy, that greatly accelerated rates of physiological aging can be driven by stress, most likely associated with lack of control over work that Wang et al. found a critical component in the etiology of AD.

The argument can be carried upward in scale.

Following Wallace and Wallace (2013), within the USA, stress and power relations between groups at the population level of organization appear particularly well-represented by two indices, percent unemployed and the percent of the workforce in a union, and we model state-level Alzheimer's disease deaths based on these variates. The dependent variate is the annual average 'young elderly' Alzheimer's death rate (ICD-10 G30 classification) per 100,000 for the age range 65–74 in the period 1999–2006 (US Centers for Disease Control data). The independent variates are the percent unemployed in 2003, and percent of the workforce union members for 2004 (US Department of Labor Statistics).

The 50-state regression model (F=7.98, P=0.0010, adjusted R^2=22.2%) is

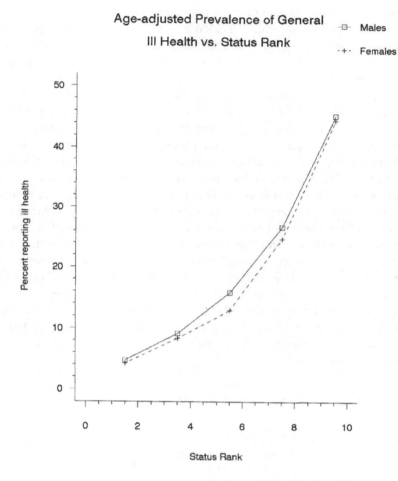

Fig. 8.2 Dose-response curves of age-adjusted self-reported ill-health vs. inverse self-reported status rank for the Whitehall II cohort of UK government workers. 1 is high, and 10 low, status.

Parameter	Estimate	SE	t	P
CONSTANT	18.76	3.86	4.86	0.0000
unemp03	1.407	0.694	2.03	0.048
unionpc04	-0.503	0.132	-3.82	0.0004

Thus union participation indexes a decrease, and unemployment an increase, of Alzheimer's mortality incidence in the young elderly, consistent with a theoretical model for which locus-of-control affects generalized inflammation, and the de facto rate of aging, long known to be the principal risk factor for Alzheimer's disease.

It is possible to further characterize these results. The Southern US states form the core of the so-called 'right-to-work' (RTW) laws that forbid requiring a worker to join a union even if employed in a work force that has union representation. RTW laws represent and constitute a culture of individualism and of active anti-collectivism. Other differences between the RTW and non-RTW states include economic history, with the former only recently industrialized and historically agricultural. The Plains and Southwestern RTW states showed the fastest rate of increase in manufacturing jobs in the 1990s, but very low rates of such jobs per unit population. Indicators of social and political engagement such as voting participation and percent employed belonging to a union showed large differences between the two groups of states. Together, they likely indicate the strength of social and political support and control within the two groups of states.

The first step in modeling AD for these two sets of states is to examine annual average rates of Alzheimer's deaths per 100,000 over the years 1999–2006 for three age cohorts: 65–74, 75–84, and 85+, and compare them using a standard t-test. This gives:

Cohort	1	2	3
RTW	23.0	182.2	843.1
Non-RTW	19.3	159.3	802.7
P(t-test)	0.02	0.04	NS

The rates, as expected, increase sharply with cohort age, but are markedly lower at all ages in the non-RTW states, and statistically significantly so in the younger.

The regression model above can be applied to all three cohorts and for the US, RTW, and non-RTW states. The percent of adjusted variance accounted for by the models, R^2, and the maximum significance P of the regressions, is as follows:

Cohort	1	2	3	Max. P
US	22.1	33.0	8.7	0.04
Non-RTW	16.6	31.8	20.8	0.04
RTW	0	12.4	0	NS

Although the RTW regressions are not significant, the others are highly so, and the raw numbers all show a peak in the middle cohort.

Figure 8.3 displays these results as a signal-to-noise ratio vs. signal amplitude graph. Age is taken as a toxic exposure, and the regression adjusted R^2 as a signal-to-noise measure for a signal transduction model.

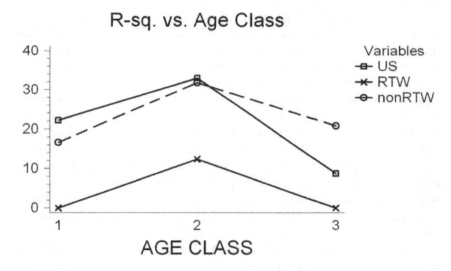

Fig. 8.3 From Wallace and Wallace 2013. Regression adjusted R^2 as signal-to-noise ratio (SNR) vs. age cohort as toxic exposure for a model based on state-level percent unemployed and percent of workforce unionized. The SNR follows a unimodal inverted U consistent with a signal transduction mechanism, with the non-RTW states shifted to higher ages. This is consistent with a perspective in which the degree of physiological meaning of the population-level response to the driving variates is changed by the underlying cultural milieu.

Two essential points are the failure of the model for the RTW states and the relative shift of the non-RTW toward the older cohort relative to the full US model. The inference is that collective efficacy lowers chronic inflammation-induced premature aging that expresses itself in Alzheimer's mortality in the young elderly, in accordance with theory. Underlying cultural context appears to profoundly affect both the rate of effective aging and the population response to patterns of affordance and stress.

One inference would be that the RTW states in particular, and the US in general, following the arguments of Heine (2001) described in the next chapter, embody a pathologically individualistic culture that is contrary to evolved human norms, a disjunction expressing itself in premature aging.

8.7   Chronic stress and therapeutic failure

Suppose therapeutic intervention against the pathological effects of chronic stress is attempted. How does ongoing stress affect treatment?

Socioculturally constructed and structured psychosocial stress, in this model having generalized grammar and syntax, can be viewed as entraining the function of zero mode identification – determining the $\mathbf{R}_0$ of Section 3.5 – most typically when the coupling with stress exceeds a threshold. More than one threshold appears likely, accounting in a sense for the typically staged nature of environmentally caused disorders. These should result in a synergistic – i.e., comorbidly excited – mixed affective disorders, and represent the effect of stress on the linked decision processes of various cognitive control and regulatory functions, in particular through the identification of a false 'zero mode' of the generalized retina (GR). This is a collective, but highly systematic, 'tuning failure' that, in the Rate Distortion sense, represents a literal image of the structure of imposed psychosocial stress written upon the ability of the GR to characterize a normal condition, causing a mixed excited state of chronically comorbid dysfunction.

In this model, different eigenmodes Y_k of the GR regression model characterized by the matrix $\mathbf{R}_0$ can be taken to represent the 'shifting-of-gears' between different 'languages' defining the sets B_0 and B_1 of Section 2.7. That is, different eigenmodes of the GR would correspond to different required (and possibly mixed) characteristic systemic responses.

If there is a state (or set of states) Y_1 such that $\mathbf{R}_0 Y_1 = Y_1$, then the 'unitary kernel' Y_1 corresponds to the condition 'no response required', the set B_0, that is, normal function.

Suppose pathology becomes manifest, i.e.,

$$\mathbf{R}_0 \rightarrow \mathbf{R}_0 + \delta\mathbf{R} \equiv \hat{\mathbf{R}}_0,$$

so that some chronic dysfunctional state becomes the new 'unitary kernel', and

$$Y_1 \to \hat{Y}_1 \neq Y_1$$

$$\hat{\mathbf{R}}_0 \hat{Y}_1 = \hat{Y}_1.$$

This could represent various forms of pathology.

Suppose we wish to induce a sequence of therapeutic counterperturbations $\delta\mathbf{T}_k$ according to the pattern

$$(\hat{\mathbf{R}}_0 + \delta\mathbf{T}_1)\hat{Y}_1 = Y^1$$
$$\hat{\mathbf{R}}_1 \equiv \hat{\mathbf{R}}_0 + \delta\mathbf{T}_1$$
$$(\hat{\mathbf{R}}_1 + \delta\mathbf{T}_2)Y^1 = Y^2$$

$$.......... \tag{8.11}$$

and so on, so that, in some sense,

$$Y^j \to Y_1 \tag{8.12}$$

That is, the system of interest, as monitored by the GR, is driven toward its original condition.

It may or may not be possible to have $\hat{\mathbf{R}}_0 \to \mathbf{R}_0$. That is, actual cure may not be possible, in which case palliation or control is the therapeutic aim.

The essential point is that the pathological state represented by $\hat{\mathbf{R}}_0$ and the sequence of therapeutic interventions $\delta\mathbf{T}_k, k = 1, 2, \ldots$ are interactive and reflective, depending on the regression of the set of vectors Y^j to the desired state Y_1, again, much in the same spirit as Jerne's immunological idiotypic hall of mirrors.

The therapeutic problem revolves around minimizing the difference between Y^k and Y_1 over the course of treatment. That difference represents the inextricable convolution of 'treatment failure' with 'adverse reactions' to the course of treatment itself, and 'failure of compliance' attributed through social construction by provider to patient, i.e., failure of the therapeutic alliance.

It should be obvious that the treatment sequence $\delta\mathbf{T}_k$ represents a cognitive path of interventions having, in turn, a dual information source in the sense we have previously invoked.

Treatment may, then, interact in the usual Rate Distortion manner with patterns of structured psychosocial stress that are, themselves, signals from an embedding information source. Thus treatment failure, adverse reactions, and patient noncompliance will, of necessity, embody a distorted image of structured psychosocial stress.

In sum, characteristic patterns of treatment failure, adverse reactions, and patient noncompliance reflecting collapse of the therapeutic alliance, will occur in virtually all therapeutic interventions according to the manner in which structured psychosocial stress is expressed as an image within the treatment process. This would most likely occur in a highly punctuated manner, depending in a quantitative way on the degree of coupling of the three-fold system of affected individual, patient/provider interaction, and treatment mode, with that stress.

Given that the principal environment of humans is defined by interaction with other humans and with enveloping socioeconomic institutions and cultural trajectories, these are likely to be very strong effects.

8.8   Implications

Cognitive phenomena pervade biology, from 'simple' wound healing, through immune maintenance and response, tumor control, neural function, social interaction, and so on. Many can be associated with 'dual' information sources constrained by the necessary conditions of information and control theories.

Cognition's ubiquity, found at every scale and level of organization of the living state, opens to evolutionary exaptation the crosstalk and noise that plague all information exchange. A principal outcome will be the repeated development of punctuated global broadcast mechanisms that can entrain sets of 'unconscious' cognitive modules into shifting, tunable cooperative arrays tasked with meeting the changing patterns of threat and opportunity that challenge all organisms. When such entrainment involves neural systems acting on a timescale of a few hundred milliseconds, the phenomenon is characterized as consciousness. When entrainment involves many individuals or cultural artifacts, the outcomes are social or institutional processes.

Most singularly, embedding high level neural global broadcasts of animal consciousness within a nested hierarchy of cognitive and other sources of information evades the logical fallacy of attributing to 'the brain' the broad spectrum of functions that can only be embodied in the full construct of the individual-in-context.

More specifically, the nested nature of biocognition for humans particularly includes sociocultural and historical context, and this enables a channel carrying a powerful downward resonance from higher to lower levels of organization. A central mechanism appears to involve pathologies from embedding and imposed psychosocial stress that drive high demand for metabolic free energy at what medical practitioners are wont to call 'basic biological' levels. That is, chronic inflammation at the cellular, tissue, and organ levels of organization within an individual can be triggered by chronic imposed stresses, most effectively those associated with power relations between groups and individuals, control over one's work so on, driving higher de-facto rates of aging, a mechanism that has broad implications for individual therapeutic and public health interventions. In particular, systematic 'meaningful' stress can not only trigger early AD and other diseases of chronic inflammation, but will most surely seriously interfere with both treatment and prevention.

Similar arguments are to be found throughout the literature. For example, as Qui et al. (2009) put it, focusing on AD,

> Alzheimer's dementia is a multifactorial disease in which older age is the strongest risk factor... [that] may partially reflect the cumulative effects of different risk and protective factors over the lifespan, including the complex interactions of genetic susceptibility, psychosocial factors, biological factors, and environmental exposures experienced over the lifespan.

They further explain that mutation effects account for only a small fraction of observed cases, and that the APOE $\epsilon4$ allele – the only established genetic factor for both early and late onset disease – is a *susceptibility* gene, neither necessary nor sufficient for disease onset. They describe how many of the same factors implicated in diabetes and cardiovascular disease predict onset of Alzheimer's as well: tobacco use, high blood pressure, high serum cholesterol, chronic inflammation, as indexed by a higher level of serum C-reactive protein, and diabetes itself. Effective protective factors include high educational and socioeconomic status, regular physical exercise, mentally demanding activities, and significant social engagement.

Qui et al. conclude that

> Epidemiological research has provided sufficient evidence that vascular risk factors in middle-aged and older adults play a significant role in the development and pro-

gression of dementia and [Alzheimer's disease], whereas extensive social network and active engagement in mental, social, and physical activities may postpone the onset of the dementing disorder. Multidomain community intervention trials are warranted to determine to what extent preventive strategies toward optimal control of multiple vascular factors and disorders, as well as the maintenance of an active lifestyle, are effective against dementia and [Alzheimer's disease].

Such considerations are likely to apply to many of the chronic diseases of aging, suggesting the need for systematic public health interventions aimed at improving living and working conditions.

Chapter 9

What is to be done?

9.1 Summary

A survey of the cultural psychology literature suggests that Western biomedicine's fascination with atomistic, individual-oriented, interventions is a culturally-specific artifact having little consonance with complex, subtle, multiscale, multilevel, social, ecological, or biological realities. Other cultural traditions may, in fact, view atomistic strategies as inherently unreal. The perspective of health promotion, by contrast, suggests that the most effective medical or public health interventions must be analogously patterned across scale and level of organization: 'magic strategies' will almost always be synergistically, and often emergently, more effective than 'magic bullets'. Multifactorial interventions focused at the human keystone ecosystem level of mesoscale social and geographic groupings may be particularly effective.

9.2 Introduction

Physiologically, mitochondrial function certainly represents a critical level of organization, and thus seems to present a leverage point in prevention and management across a broad spectrum of chronic diseases. For example, Yao et al. (2009) write

> [M]itochondrial dysfunction is more than the sum of
> its components. Sustained disturbances in the pathway
> balances of functional efficiency can lead to systems-level
> defects not observable in additional individual components
> until later. Such early imbalances can lead to accumulated

changes in the mitochondria that later emerge as observable changes in other aspects of the system... Consistent with mitochondrial dysfunction, decreased mitochondrial bioenergetics has been demonstrated to cause amyloid production and nerve cell atrophy... A direct link between $A\beta$-induced toxicity and mitochondrial dysfunction in AD has been suggested... If mitochondrial dysfunction is a causal link to Alzheimer's, the susceptibility of mitochondria to environmental and genetic risk factors should be a critical factor in the development of late onset sporadic AD...

They conclude that "...[A range of] studies indicate a therapeutic strategy to prevent AD by sustaining mitochondrial metabolic function".
Similarly, Beal (2007) argues

There is increasing evidence linking mitochondrial dysfunction to neurodegenerative diseases. Mitochondria are critical regulators of cell death, a key feature of neurodegeneration... This is the case in Alzheimer's disease, in which there is evidence that both β-amyloid and the amyloid precursor protein may directly interact with mitochondria, leading to increased free radical production... [In addition] an impressive number of disease specific proteins interact with mitochondria. Therapies that target basic mitochondrial processes such as energy metabolism in free radical generation, or specific interactions of disease-related protein with mitochondria, hold great promise.

Writing in the pharmaceutical literature, Johri and Beal (2012) conclude

There are... a number of promising new compounds and therapeutic targets that modulate mitochondria and produce neuroprotective effects... These compounds show great promise for treating patients who suffer from neurodegenerative diseases, for which there is as yet no effective treatment to slow or halt the underlying disease processes.

Unfortunately, a well-defined physiological leverage point does not necessarily imply the possibility of a therapeutic magic bullet, in the sense of contemporary Western medicine. Since mitochondrial energy dynamics are so central to life, it seems likely that direct interventions – small-molecule pharmaceuticals or targeted nutraceuticals – would risk a plethora of very nasty 'side effects', likely triggering an iatrogenic version of the generalized inflammation described in the last chapter.

At the least, these considerations suggest, from the beginning, the need to aim for combination therapies – multiple, simultaneous, 'small molecule interventions' – to accurately target any magic bullet mitochondrial boost, very carefully avoiding 'shotgun' impacts.

More generally, however, but no less centrally, two broad conundrums currently challenge the Western biomedical paradigm, and these have implications for intervention strategies aimed at improving mitochondrial function. The first, shown in figure 9.1, adapted from Bernstein (2010), involves an 'inverse Moore's Law' for pharmaceuticals. The figure shows the number of USFDA approvals per inflation-adjusted $ billion research investment (2010 dollars), 1950-2010. The log-linear 'decline in research productivity' represents the failure of complex physiological processes to respond to simple interventions, and, in a real sense, traces the failure of molecular biology and other reductionist approaches to cash out on their considerable intellectual and financial investments. Current pharmaceutical industry strategies involve the retreat into large-scale marketing fraud, rampant consolidation and consequent declines in research, and financial engineering schemes such as off-shore tax reduction (Wallace and Wallace 2013). Biological engineering approaches to drug development, while increasingly pursued, have usually resulted in exorbitantly expensive products.

Figure 8.2 characterizes the other conundrum. As described, it shows sex-specific dose-response curves of age-adjusted self-reported ill-health vs. an inverse self-reported status rank for the Whitehall II cohort of privileged UK government workers. In sum, accelerated rates of physiological aging can be driven by lack of control over work.

As Wallace and Wallace (2013) argue at some length, successful health interventions under such conditions must act synergistically across scale and level of organization, addressing not only some particular 'molecular pathology', but, in this circumstance, actively reducing and/or buffering, the psychosocial stresses that accelerate physiological aging.

An inference from this example is that effective therapy against more explicitly 'genetic' mitochondrial dysfunctions may similarly require synergistic interventions at more than a single scale or level of organization. That is, 'magic bullets' are primarily a Western cultural conceit (Wallace and Wallace 2013), and figure 9.1 suggests their effectiveness is waning.

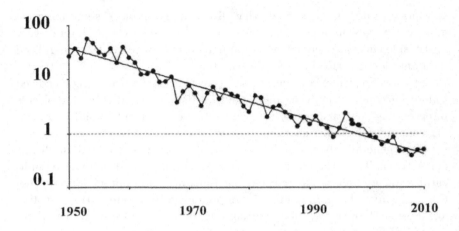

Fig. 9.1   Inverse Moore's Law for the pharmaceutical industry: Inflation-adjusted (2010 $) number of USFDA approvals per $ billion research investment, 1950-2010. This represents, among other dynamics, the failure of molecular biology to cash out its considerable intellectual and financial investment.

What are some alternatives?

9.3   Health promotion

A metareview by Jackson et al. (2007) identifies a number of factors essential to successful health promotion strategies that go beyond the individual-oriented medical treatments characterized by figure 9.1:

1. Investment in building healthy public policy is a key strategy

Relevant public actions include investment in government and social policy, the creation of legislation and regulations and intersectoral and interorganizational partnerships and collaboration. In some cases healthy public policy was the strategy for which the most evidence of effectiveness exists: e.g., legislation for road safety and social policy for income security and poverty reduction.

2. Supportive environments need to be created at all levels

Such efforts include a variety of actions that represent supportive conditions at the structural (policy), social (including community) and individual levels. Key activities might include providing instrumental supports such as food vouchers or supplements, group support, nutritional education, counseling and home visits. Supportive environments are required for success at all levels of health promotion strategies.

3. Effectiveness of community action is unclear and requires further evidence

Although the impact of such actions in terms of behavior change has ranged from modest to disappointing, they have achieved success in terms of community and systems change.

4. Personal skills development must be combined with other strategies for effectiveness

Health education and related strategies were ineffective if implemented in isolation from other strategies that create structural-level conditions to support health and increase access to goods, products, and services. Intervention must address not only the health issues, but also the social and economic conditions that lead to risk behaviors.

5. Interventions employing multiple strategies and actions at multiple levels and sectors are most effective

The most effective interventions employ multiple health promotion strategies, operate at multiple levels (often including all of the structural, social group and personal levels), work in partnership across sectors and include a combination of integrated actions to support each strategy. Noncommunicable disease interventions in particular must employ multiple strategies and actions at multiple levels. Schools, workplaces and municipalities were found to be effective settings for many interventions because they provide opportunities to effectively reach large numbers of people with sustained interventions.

Among the important conclusions Jackson et al. reached was the observation that health promotion interventions are only effective when relevant to the context in which they are being used, including awareness of the social, cultural, economic and political contexts. The goals, strategies and actions of any intervention must be relevant and appropriate to the people they aim to reach and the systems they aim to work within.

In the previous chapter we outlined the 'basic biology' that supports these observations, understanding that for humans, an utterly essential environmental context is defined by relations, not only with other individuals, but, as illustrated by figure 8.2, by the embedding power relations between groups.

9.4   Western cultural atomism

Why does the very idea that magic bullets might be less effective than more comprehensive multilevel strategies seem so inherently alien in the context of Western biomedicine? Why, in the face of the inverse Moore's Law that has driven the catastrophic collapse of pharmaceutical industry productivity, shown in figure 9.1, has the general response been one of Translational Medicine, i.e., more-of-the-same-but-better? Susser (1973), in his famous book *Causal Thinking in the Health Sciences*, some time ago explored the inadequacies of atomistic thinking in the study of health and illness. As Kaufman and Poole (2000) note, however, little has really changed in the basic ideology of the field, arguing that although a recent resurgence of interest in social context has revivified many of the points made by Susser in 1973, the formalization of this ecologic perspective unfortunately advanced little in the subsequent quarter-century. The progress toward more refined and systematic articulation of causal logic in the epidemiological and statistical literature in recent decades has, in their view, been characterized by an explicit conceptual foundation in atomistic interventions. The emergent properties of causal systems, as distinct from the consideration of multiple discrete actions, remains largely undescribed in any formal sense in the epidemiological literature.

Even such a supposedly multifactorial approach as network-based systems biology (e.g., Pujol et al. 2009; Boran and Iyengar 2010; Dudley et al. 2010; Arrell and Terzic, 2010) is focused primarily on drug development for individual-level treatment, with the 'system' outlined in a major review article (Zhao and Iyengar, 2012, figure 1) as:

Atomic/molecular interactions → Cellular/tissue-level networking and physiology → Organ-level networking and physiology → Whole-body outcomes.

Similar conceptual failures, in fact, plague economics and evolutionary theory. That is, the search for magic bullets – atomized causality – is ubiquitous in Western thought. This is, however, a social construct – a conceptual artifact – not shared by other cultures, and the resulting dissonance may singularly affect the success or failure of collaborations with partners having East Asian acculturation.

Why?

Tony Lawson's (2006) examination of heterodox economics serves as a counterintuitive starting point. It focuses, first, on characterizing the essential features of the mainstream tradition in Western economic theory as involving explicitly physics-like, deductive mathematical models of social phenomena that inherently require an atomistic perspective on individual, isolated economic actors, a methodology subject to scathing commentary at the highest levels (e.g., Leontief, 1982). Lawson (2006) characterizes the need for isolated atomism in mainstream theory as arising from mathematical considerations, in that deductivist theorizing of the sort pursued in modern economics ultimately has to be couched in terms of such 'atoms' just to ensure that under conditions x the same (predictable or deducible) outcome y always follows. The point then, from his perspective, is that the ontological presuppositions of the insistence on mathematical modeling include the restriction that the social domain is everywhere constituted by sets of isolated atoms.

A converse interpretation, however, is also possible: that cultural assumptions of atomicity can drive the particular forms of mathematical models chosen by researchers.

Lawson (2006) describes various heterodox economic approaches – post Keynsianism, (old) institutionalism, feminist, social, Marxian, Austrian and social economics, among others – as representing something of a generalized social science in which the dominant emphases of the separate heterodox traditions are just manifestations of categories of social reality that conflict with the assumption that social life is everywhere composed of isolated atoms.

More recently, criticism has emerged of gene-based replicator dynamics versions of evolutionary theory that suffer similar atomistic model constrictions (e.g., Lewontin 2000, 2010). Much of the debate in evolutionary theory has revolved around the 'basic' target of selection, with the Modern Evolutionary Synthesis heavily invested in the atomistic, gradualist theory of mathematical population genetics (e.g., Ewens, 2004). Heterodox, non-atomistic, heavily contextual, evolutionary theories have emerged that materially challenge and extend that Synthesis (e.g., Gould 2002, Odling-Smee et al. 2003, Wallace 2010).

Economics and evolutionary theory, however, are not the only biological/social sciences to come under the same gun. The cultural psychologist S. Heine (2001) finds that the extreme nature of American individualism suggests that a psychology based on late 20th century American research not only stands the risk of developing models that are particular to that culture, but of developing an understanding of the self that is peculiar in the context of the world's cultures.

The explanation for this pattern goes deeper than ideology, into the very bones of Western culture: Nisbett et al. (2001), following in a long line of research (Markus and Kitayama 1991, and the summary by Heine 2001), review an extensive literature on empirical studies of basic cognitive differences between individuals raised in what they call 'East Asian' and 'Western' cultural heritages, which they characterize, respectively, as 'holistic' and 'analytic'. They argue:

1. Social organization directs attention to some aspects of the perceptual field at the expense of others.

2. What is attended to influences metaphysics.

3. Metaphysics guides tacit epistemology, that is, beliefs about the nature of the world and causality.

4. Epistemology dictates the development and application of some cognitive processes at the expense of others.

5. Social organization can directly affect the plausibility of metaphysical assumptions, such as whether causality should be regarded as residing in the field vs. in the object.

6. Social organization and social practice can directly influence the development and use of cognitive processes such as dialectical vs. logical ones.

Nisbett et al. (2001) conclude that tools of thought embody a culture's intellectual history, that tools have theories built into them, and that users accept these theories, albeit unknowingly, when they use these tools.

Masuda and Nisbett (2006) find research on perception and cognition suggesting that, whereas East Asians view the world holistically, attending to the entire field and relations among objects, Westerners view the world analytically, focusing on the attributes of salient objects. Compared to Americans, East Asians were more sensitive to contextual changes than to focal object changes. These results suggest that there can be cultural variation in what may seem to be basic perceptual processes.

Nisbett and Miyamoto (2005) similarly found evidence that perceptual processes are influenced by culture.

Wallace (2007) argues that a necessary conditions mathematical treatment of Baars's global workspace consciousness model, analogous to Dretske's communication theory analysis of high level mental function, can be used to explore the effects of embedding cultural heritage on inattentional blindness. Culture should express itself in this basic psychophysical phenomenon across a great variety of sensory modalities because conscious attention must conform to constraints generated by cultural context.

In sum, profound, culturally-based, atomistic ideological constraints abound across a plethora of Western scientific disciplines, including contemporary biomedicine. Engaging other cultural sensibilities not so constrained, particularly on matters of health and illness, will almost surely require adoption of a magic strategy perspective. This is, first, because such strategies are likely to work better in the long run than atomistic interventions, and second, because East Asian and other potential markets may respond better to such approaches than to being force-fed what is widely understood (by them) to be a Western cultural conceit.

Indeed, the previous chapter reexamined and reinterpreted a nested set of biological and other broadly inflammatory modules from the perspective of Chapter 2, providing new insights on multiscale, multilevel health interventions. It characterized both individual cognitive processes and punctuated correlations of them that recruit sets of cognitive modules into a larger whole acting across scale or level of organization. The various examples are much in the spirit of Maturana and Varela (1980, 1992) who long ago understood the central role that cognition must play in biological phenomena. The chronic 'inflammatory' activation of individual cognitive modules and their larger working coalitions seems of interest, as well as the excess demand for metabolic free energy such activation implies, leading to possible synergistic dysfunctional dynamics.

9.5   Implications

We have described how Western and East Asian cultural heritage affects fundamental perceptual mechanisms. What had been until recently considered basic biological phenomena prove to be greatly modulated, indeed, inverted, by cultural influence: Westerners focus on objects atomized from their context, while East Asians focus on the context itself.

Given that surprising result – at least surprising to a certain class of Western scientists, if not to anthropologists – it seems wise to ask further questions regarding the role of culture in the 'basic biology' of HPA axis, blood pressure, protein folding, tumorigenesis, mental and developmental disorders, and many other pathologies, particularly in the context of ongoing psychosocial stress.

Recall Heine's (2001) caution that the extreme nature of American individualism implies that a psychology based on 20th Century American research runs the risk of becoming an ideological *ignis fatuus*, a Western cultural artifact divorced from reality. This may, in fact, be an example that generalizes across the study of many broadly cognitive phenomena. In particular, the definition and dynamic impact of stress may be culturally contingent, in terms of overt manifestation, progression, and individual and collective experience.

Contrary to Western medicine's assumptions regarding 'basic biology', for much of human health and illness, one size may not fit all. Indeed, even malaria among cohabiting peoples in Burkina Faso seems much the different disease for former masters and former slaves via an immuno-cultural construct (Wallace and Wallace 2002).

Indeed, Global Workspace Theory (Baars 1988; Baars et al. 2013) makes explicit reference to contexts/frames that channel conscious experience, forming, in effect, the riverbanks that channel the classic 'stream of consciousness'. It is clear that virtually all biological broadcast phenomena are similarly channeled, and that, for humans, culture must be a principal determinant of such framing. To reiterate, as the evolutionary anthropologist Robert Boyd put it, 'culture is as much a part of human biology as the enamel on our teeth' (e.g., Richerson and Boyd 2006).

Incorporating this mechanism to explicitly include the synergism between inflammatory activation and mitochondrial dysfunction emerges using the results of Chapter 6.

The symmetry breaking arguments of Chapter 2, leading to Section 2.9 in which average distortion is interpreted as an order parameter, make an important basic case: metabolic free energy, largely delivered as ATP through mitochondrial dynamics, is a central linchpin for the biological automata that interact to constitute the living state. Failures to deliver adequate rates of that resource will express themselves in the often punctuated staging of culturally-modulated chronic disease, according to this model.

More explicitly, using the arguments of Chapter 3, the metabolic cost of physiological regulation can grow at rates at least proportional to the required regulatory Rate Distortion Function, a convex function of the average distortion between regulatory intent and effect. The constants of proportionality κ_1, κ_2 in equations (3.7) and (3.15), may be very large, given the usually massive entropic losses associated with biological process. This, in conjunction with the arguments above, suggests that exposure to noxious chemical agents, infections, culturally-meaningful psychosocial stress, and so on, can act as a kind of noise, raising values of β and σ in equation (3.15), triggering transitions to chronic, quasi-stable, high levels of distortion. This causes massive consumption of metabolic free energy – generalized inflammation leading to early onset of chronic disease. Therapeutic intervention must overcome the quasi-stability of these pathological states. Given the path-dependent nature of physiological development, a return to previous modes may be impossible, requiring ongoing treatment that may be profoundly affected by ongoing stressors.

A central consequence of the crosstalk underlying figure 2.3 is that there is unlikely to be much in the way of 'simple' generalized inflammatory chronic disease. That is, serious comorbidity – perturbations can resonate across the full set of bioregulatory systems – are not only inevitable, but may often be an unfortunate consequence of therapeutic intervention as well. Thus, synergistic pairing of medical with appropriate neighborhood/social network interventions – for humans, a keystone level of organization in Holling's (1992) sense – would be expected to (Wallace and Wallace 2004):

(1) Damp down unwanted treatment side effects.

(2) Make the therapeutic alliance between practitioner and patient more effective.

(3) Improve patient compliance.

(4) Enhance placebo effect.

(5) In the context of real stress reduction, synergistically improve the actual biological impacts of medical interventions or prevention strategies.

Even without pathogenic events or exposures, normal aging may make it impossible to provide rates of metabolic free energy needed for routine regulation, even in the absence of noxious agencies, leading to increasing distortion in ordinary physiological activities – systematic degradation of the organism – causing the spontaneous phenotypic shifts that constitute senescence. The triggering of shifts between quasi-stable system modes by external perturbations leading to generalized inflammation may, in fact, represent premature senescence, from this perspective. Culturally-appropriate cross-scale interventions – magic strategies – might well slow, or even, in some measure, reverse the effects of, such mechanisms.

While, as Jackson et al. (2007) point out, improvements in living and working conditions – what they characterize as 'healthy public policy' – has the most direct impact, reaching traditionally isolated, vulnerable populations trapped in a pathological historical trajectory, or any population in the absence of healthy public policy, will require special focus on keystone social network and geographic mesoscales.

Chapter 10

Mathematical Appendix

10.1 Morse Theory

The basic idea of Morse Theory is to examine an n-dimensional manifold M as decomposed into level sets of some function $f : M \to \mathbf{R}$ where $\mathbf{R}$ is the set of real numbers. The a-level set of f is defined as

$$f^{-1}(a) = \{x \in M : f(x) = a\},$$

the set of all points in M with $f(x) = a$. If M is compact, then the whole manifold can be decomposed into such slices in a canonical fashion between two limits, defined by the minimum and maximum of f on M. Let the part of M below a be defined as

$$M_a = f^{-1}(-\infty, a] = \{x \in M : f(x) \le a\}.$$

These sets describe the whole manifold as a varies between the minimum and maximum of f.

Morse functions are defined as a particular set of smooth functions $f : M \to \mathbf{R}$ as follows. Suppose a function f has a critical point x_c, so that the derivative $df(x_c) = 0$, with critical value $f(x_c)$. Then, f is a Morse function if its critical points are nondegenerate in the sense that the Hessian matrix of second derivatives at x_c, whose elements, in terms of local coordinates are

$$\mathcal{H}_{i,j} = \partial^2 f / \partial x^i \partial x^j,$$

has rank n, which means that it has only nonzero eigenvalues, so that there are no lines or surfaces of critical points and, ultimately, critical points are isolated.

The index of the critical point is the number of negative eigenvalues of $\mathcal{H}$ at x_c.

A level set $f^{-1}(a)$ of f is called a critical level if a is a critical value of f, that is, if there is at least one critical point $x_c \in f^{-1}(a)$.

Following Pettini (2007), the essential results of Morse Theory are:

1. If an interval $[a, b]$ contains no critical values of f, then the topology of $f^{-1}[a, v]$ does not change for any $v \in (a, b]$. Importantly, the result is valid even if f is not a Morse function, but only a smooth function.

2. If the interval $[a, b]$ contains critical values, the topology of $f^{-1}[a, v]$ changes in a manner determined by the properties of the matrix H at the critical points.

3. If $f : M \to \mathbf{R}$ is a Morse function, the set of all the critical points of f is a discrete subset of M, i.e., critical points are isolated. This is Sard's Theorem.

4. If $f : M \to \mathbf{R}$ is a Morse function, with M compact, then on a finite interval $[a, b] \subset \mathbf{R}$, there is only a finite number of critical points p of f such that $f(p) \in [a, b]$. The set of critical values of f is a discrete set of $\mathbf{R}$.

5. For any differentiable manifold M, the set of Morse functions on M is an open dense set in the set of real functions of M of differentiability class r for $0 \leq r \leq \infty$.

6. Some topological invariants of M, that is, quantities that are the same for all the manifolds that have the same topology as M, can be estimated and sometimes computed exactly once all the critical points of f are known: let the Morse numbers $\mu_i(i = 0, ..., m)$ of a function f on M be the number of critical points of f of index i, (the number of negative eigenvalues of $\mathcal{H}$). The Euler characteristic of the complicated manifold M can be expressed as the alternating sum of the Morse numbers of any Morse function on M,

$$\chi = \sum_{i=1}^{m} (-1)^i \mu_i.$$

The Euler characteristic reduces, in the case of a simple polyhedron, to

$$\chi = V - E + F$$

where V, E, and F are the numbers of vertices, edges, and faces in the polyhedron.

7. Another important theorem states that, if the interval $[a, b]$ contains a critical value of f with a single critical point x_c, then the topology of the set M_b defined above differs from that of M_a in a way which is determined by the index, i, of the critical point. Then M_b is homeomorphic to the manifold obtained from attaching to M_a an i-handle, i.e., the direct product of an i-disk and an $(m - i)$-disk.

Pettini (2007) and Matsumoto (2002) contain details and further references.

10.2 Groupoids

A groupoid, G, is defined by a base set A upon which some mapping – a morphism – can be defined. Note that not all possible pairs of states (a_j, a_k) in the base set A can be connected by such a morphism. Those that can define the groupoid element, a morphism $g = (a_j, a_k)$ having the natural inverse $g^{-1} = (a_k, a_j)$. Given such a pairing, it is possible to define 'natural' end-point maps $\alpha(g) = a_j, \beta(g) = a_k$ from the set of morphisms G into A, and a formally associative product in the groupoid $g_1 g_2$ provided $\alpha(g_1 g_2) = \alpha(g_1), \beta(g_1 g_2) = \beta(g_2)$, and $\beta(g_1) = \alpha(g_2)$. Then, the product is defined, and associative, $(g_1 g_2) g_3 = g_1(g_2 g_3)$. In addition, there are natural left and right identity elements λ_g, ρ_g such that $\lambda_g g = g = g \rho_g$.

An orbit of the groupoid G over A is an equivalence class for the relation $a_j \sim G a_k$ if and only if there is a groupoid element g with $\alpha(g) = a_j$ and $\beta(g) = a_k$. A groupoid is called transitive if it has just one orbit. The transitive groupoids are the building blocks of groupoids in that there is a natural decomposition of the base space of a general groupoid into orbits. Over each orbit there is a transitive groupoid, and the disjoint union of these transitive groupoids is the original groupoid. Conversely, the disjoint union of groupoids is itself a groupoid.

The isotropy group of $a \in X$ consists of those g in G with $\alpha(g) = a = \beta(g)$. These groups prove fundamental to classifying groupoids.

If G is any groupoid over A, the map $(\alpha, \beta) : G \to A \times A$ is a morphism from G to the pair groupoid of A. The image of (α, β) is the orbit equivalence relation $\sim G$, and the functional kernel is the union of the isotropy groups. If $f : X \to Y$ is a function, then the kernel of f, $ker(f) = [(x_1, x_2) \in X \times X : f(x_1) = f(x_2)]$ defines an equivalence relation.

Groupoids may have additional structure. For example, a groupoid G is a topological groupoid over a base space X if G and X are topological spaces and α, β and multiplication are continuous maps.

In essence, a groupoid is a category in which all morphisms have an inverse, here defined in terms of connection to a base point by a meaningful path of an information source dual to a cognitive process.

The morphism (α, β) suggests another way of looking at groupoids. A groupoid over A identifies not only which elements of A are equivalent to one another (isomorphic), but *it also parameterizes the different ways (isomorphisms) in which two elements can be equivalent*, i.e., in our context, all possible information sources dual to some cognitive process. Given the information theoretic characterization of cognition presented above, this produces a full modular cognitive network in a highly natural manner.

10.3   Deterministic finite state automata

A deterministic finite automaton (Kozen, 1997) is represented by a quintuple (Q, Σ, h, q_0, F) where:

Q is a finite set of states.

Σ is a finite set of symbols, the alphabet of the automaton.

h is the transition function $h : Q \times \Sigma \to Q$.

$q_0 \in Q$ is the start state.

F is a set of states of Q ($F \subseteq Q$) called accept states.

An automaton reads a finite string of symbols $a_1, ..., a_n$, where $a_i \in \Sigma$, called in input word, with the set of all words denoted by Σ^*.

A sequence of states $q_0, q_1, ..., q_n$, where $q_i \in Q$, such that q_0 is the start state and $q_i = h(q_{i-1}, a_i)$ for $0 < i \leq n$, is a run of the automaton on an input word $w = a_1, a_2, ..., a_n \in \Sigma^*$, where q_n is the final state of the run.

A word $w \in \Sigma^*$ is accepted by the automaton if $q_n \in F$.

An automaton can recognize a formal language. The language $L \subseteq \Sigma^*$ recognized by an automaton is the set of all words that are accepted by the automaton. The recognizable languages are a set of languages that are recognized by some automaton. For different automata, the recognizable languages are different. In particular, nondeterministic pushdown automata that recognize context-free languages are characterized by a more complicated 7-tuple structure.

It should be obvious that one can express a subset of the conditions in which the 'tuning theorem' of Chapter 1 applies in terms of finite automata, as well as the tuning implicit in figure 2.3.

Bibliography

Adami, C., 2012, The use of information theory in evolutionary biology, Annals of the New York Academy of Sciences, 1256:49-65.

Aguzzi, A., 2014, Alzheimer's disease under strain, Nature, 512:32-33.

Albert, R, A. Barabasi, 2002, Statistical mechanics of complex networks, Reviews of Modern Physics, 74:47-97.

Anfinsen, C., 1973, Principles that govern the folding of protein chains, Science, 181:223-230.

Arrell, D., A. Terzic, 2010, Network systems biology for drug discovery, Clinical Pharmacology and Therapeutics, 88:120-125.

Atiyah, M., I. Singer, 1963, The index of elliptical operators on compact manifolds, Bulletin of the American Mathematical Society, 69:322-433.

Atlan, H., I. Cohen, I., 1998, Immune information, self-organization, and meaning, International Immunology, 10:711-717.

Atmanspacher, H., 2006, Toward an information theoretical implementation of contextual conditions for consciousness, Acta Biotheoretica, 54:157-160.

Avital, E., E. Jablonka, 2000, Animal Traditions: Behavioral Inheritance in Evolution, Cambridge University Press, New York.

Baars, B., 1988, A Cognitive Theory of Consciousness, Cambridge University Press, New York.

Baars, B., 2005, Global workspace theory of consciousness: toward a cognitive neuroscience of human experience, Progress in Brain Research, 150:45-53.

Baars, B., S. Franklin, 2003, How conscious experience and working memory interact, Trends in Cognitive Science, 7:166-172.

Baars, B., S. Franklin, T.Z. Ramsoy, 2013, Global workspace dynamics: cortical 'binding and propagation' enables conscious contents, Frontiers in Psychology, 4:Article 200.

Backdahl, L., A. Bushell, S.Beck, 2009, Inflammatory signallilng as mediator of epigenetic modulation in tissue-specific chronic inflammation, International Journal of Biochemistry and Cell Biology, 41:176-184.

Baianu, I. et al., 2005 Complex nonlinear biodynamics in categories: higher dimensional algebra, and Lukasiewicz-Moisil Topos: transformations of neuronal, genetic and neoplastic networks, Axiomathes, 16:65-122.

Balasubramanian, K., 1980, The symmetry groups of nonrigid molecules as generalized wreath products and their representations, Journal of Chemical Physics, 72:665-677.

Barrett, L., P. Henzi, 2005, The social nature of primate cognition, Proceedings of the Royal Society B, 272:1865-1875.

Beal, M.F., 2007, Mitochondria and neurodegeneration. In Mitochondrial biology: new perspectives, Novartis Foundation Symposium 287:183-196.

Beaudry, M., F. Lemieux, D. Therien, 2005, Groupoids that recognize only regular languages. In Caires et al. (Eds.). ICALP 2005, LNCS3580: 421-433, Springer, New York.

Beck, C., F. Schlogl, 1995, Thermodynamics of Chaotic Systems, Cambridge University Press, New York.

Bennett, C., 1988, Logical depth and physical complexity. In The Universal Turing Machine: A Half-Century Survey, Herkin R. (ed.) pp. 227-257, Oxford University Press, New York.

Ben-Shachar, D., 2002, Mitochondrial dysfunction in schizophrenia: a possible linkage to dopamine, Journal of Neurochemistry, 83:1241-1251.

Bernstein Research, 2010, The Long View - R & D Productivity, Boston, MA.

Bingham, G., 1988, Task-specific devices and the perceptual bottleneck, Human Movement Science, 7:225-264.

Binney, J., N. Dowrick, A. Fisher, M. Newman, 1986, The Theory of Critical Phenomena, Clarendon Press, Oxford, UK.

Black, F., M. Scholes, 1973, The pricing of options and corporate liabilities, Journal of Political Economy, 81:637-654.

Blagosklonny, M., 2010, Why the disposable soma theory cannot explain why women live longer and why we age, Aging, 2:884-887.

Bjorntorp, P., 2001, Do stress reactions cause abdominal obesity and cormorbidities? Obesity Reviews, 2:73-86.

Boran, A., R. Iyengar, 2010, Systems approaches to polypharmacology and drug discovery, Current Opinion in Drug Discovery Development, 13:297-309.

Bos, R., 2007, Continuous representations of groupoids, arXiv:math/0612639.

Bosma-Den Boer, M., M. Van Wetten, L. Pruimboom, 2012, Chronic inflammatory diseases are stimulated by current lifestyle: how diet, stress levels and medication prevent our body from recovering, BMC Nutrition and Metabolism, 9:32.

Boyd, R., P. Richerson, 2005, The Origin and Evolution of Cultures, Oxford University Press, New York.

Brennan, R. Cheong, A. Levchenko, 2012, How information theory handles cell signaling and uncertainty, Science, 338:334-335.

Brown, R., 1987, From groups to groupoids: a brief survey, Bulletin of the London Mathematical Society 19:113-134.

Brown, R. et al., 2011, Nonabelian Algebraic Topology: Filtered Spaces, Crossed Complexes, Cubical Homotopy Groupoids, EMS Tracts in Mathematics Vol. 15.

Bruce, M., B. Beech, M. Sims, T. Brown, S. Wyatt, H. Taylor, D. Williams,

E. Crook, 2009, Social environmental stressors, psychological factors, and kidney disease, Journal of Investigative Medicine, 57:583-589.

Budrikis, Z., *et al.*, 2014, Protein accumulation in the endoplasmic reticulum as a non-equilibrium phase transition, nature:communications 5:3620.

Buneci, M., 2003, Representare de Groupoizi, Editura Mirton, Timosoara.

Byers, N., 1999, Noether's discovery of the deep connection between symmetries and conservation laws, Israel Mathematical Conference Proceedings, Vol. 12, ArXiv physics/9807044.

Champagnat, N., R. Ferriere ,S. Meleard, 2006, Unifying evolutionary dynamics: from individual stochastic process to macroscopic models, Theoretical Population Biology, 69:297-321.

Chou, K.C. G. Maggiora, 1998 Domain structural class prediction, Protein Engineering, 11:523-528.

Clark, A., 1998, Embodied, situated and distributed cognition. In Bechtal, W., G. Graham (eds.), A Companion to Cognitive Science, pp. 506-517, Blackwell, London.

Clay, H., S. Sillivan, C. Konradi, 2011, Mitochondrial dysfunction and pathology in bipolar disorder and schizophrenia, International Journal of Developmental Neuroscience, 29:311-324.

Cobb, N., F. Sonnichsen, H. Mchaourab, W. Surevicz, 2007, Molecular architecture of human prion protein amyloid: a parallel, in-register β-structure, PNAS, 104:18946-18951.

Cohen, I., 1992, The cognitive principle challenges clonal selection, Immunology Today, 13:441-444.

Cohen, I., 2000, Tending Adam's Garden: evolving the cognitive immune self, Academic Press, New York.

Cohen, I., 2006, Immune system computation and the immunological hommunculus. In Nierstrasz, O., J. Whittle, D. harel, G. Reddio (eds.), MoDels 2006, LNCS, vol. 4199, pp. 499-512, Springer, Heidelberg.

Cohen, I., D. Harel, 2007, Explaining a complex living system: dynamics, multiscaling, and emergence, Journal of The Royal Society: Interface, 4:175-182.

Cohen, S., D. Janicki-Deverts, G. Miller, 2007, Psychological stress and disease, Journal of the American Medical Association, 298:1685-1687.

Cohen, S., D. Janicki-Deverts, W. Doyle G. Miller, E. Frank, B. Rabin, R. Turner, 2012, Chronic stress, glucocorticoid receptor resistance, inflammation, and disease risk, Proceedings of the National Academy of Sciences, doi/10.1073/pnas.1118355109.

Collinge, J., A. Clarke, 2007, A general model of prion strains and their pathogenicity, Science, 318:930-936.

Cover, T., J. Thomas, 2006, Elements of Information Theory, Second Edition, Wiley, New York.

Crowell, R., R. Fox, 1963, Introduction to Knot Theory, Ginn and Company, New York.

Crowson, C., K. Liang, T. Therneau, H. Kremers, S. Gabriel, 2010, Could accelerated aging explain excess mortality in patients with seropositive rheumatoid arthritis? Arthritis and Rheumatism, 62:378-382.

Csizmok, V., P. Tompa, 2009, Structural disorder and its connection with misfolding diseases. Ch. 1 in Ovadi J., F. Orosz (eds.) Protein Folding and Misfolding: Neurodegenerative Diseases, Focus on Structural Biology, 7, Springer, New York.

Dam, T., T. Gerken, B. Cavada. et al., 2007, Binding studies of α-GalNAc-specific lectins to the α-GalNAc(Tn-antigen) form of procine submaxilary mucin and its smaller fragments, Journal of Biological Chemistry, 38:28256-28263.

Dam, T., C. Brewer, 2008, Effects of clustered epitopes in multivalent ligan-receptor interactions, Biochemistry, 47:8470-8476.

Dam, T., C. Brewer, 2010, Lectins as pattern recognition molecules: the effects of epitope density in innate immunity Glycobiology, 20:270-279.

Dayan, P., N. Daw, 2008, Decision theory, reinforcement learning, and the brain, Cognitive, Affective, and Behavioral Neuroscience, 8:429-453.

de Groot, S., P. Mazur, 1984, Nonequilibrium Thermodynamics, Dover, New York.

Dembo, A., O. Zeitouni, 1998, Large Deviations and Applications, Springer, New York.

Derrida, B., 2007, Nonequilibrium steady states, Journal of Statistical Mechanics: Theory and Experiment,
1742-5468/07/P07023.

Desjarlais, R. et al., 1995, World Mental Health, Oxford University Press, New York.

Dill, K., S. Banu Ozkan, T. Weikl, J. Chodera, V. Voelz, 2007, The protein folding problem: when will it be solved? Current Opinion in Structural Biology, 17:342-346.

Dobson, C., 2003, Protein folding and misfolding, Nature, 426:884-890.

Dretske, F., 1994, The explanatory role of information, Philosophical Transactions of the Royal Society A, 349:59-70.

Dudley, J., E. Schadt, M. Sirota, A. Butte, E. Ashley, 2010, Drug discovery in a multidimensional world: systems, patterns, and networks, Journal of Cardiovascular Translational Research, 3:438-447.

Durham, W., 1991, Coevolution: Genes, Culture, and Human Diversity, Stanford University Press, Palo Alto.

Elger, C., K. Lenertz, 1998, Seizure prediction by non-linear time series analysis of brain electrical activity, European Journal of Neuroscience, 10:786-789.

Ellis, R., 1985, Entropy, Large Deviations, and Statistical Mechanics, Springer, New York.

Epel, E., E. Blackburn ,J. Lin, F. Dhabhar, N. Adler, J. Morrow, R. Cawthon, 2004, Proceedings of the National Academy of Sciences, 101:17312-17315.

Ewens, W., 2004, Mathematical Population Genetics, Springer, New York.

Falsig, J., K. Nilsson, T. Knowles, A. Aguzzi, 2008, Chemical and biophysical insights into the propagation of prion strains HFSP Journal 2:332-341.

Feynman, R., 2000, Lectures on Computation, Westview Press, New York.

Foley, D.,J. Craig, R. Morley, C. Olsson, T. Dwyer, K. Smith et al., 2009,

Prospects for epigenetic epidemiology, American Journal of Epidemiology, 169:389-400.

Forlenza, M., A. Baum, 2000, Psychosocial influences on cancer progression: alternative cellular and molecular mechanisms, Current Opinion in Psychiatry, 13:639-645.

Friedland, R., E. Koss, J. Haxby et al., 1988, Alzheimer diseae: clinical and biological heterogeniety, Annals of Internal Medicine, 109:298-311.

Friston, K., 2010, The free energy principle: a unified brain theory? Nature Reviews Neuroscience 11:127-138.

Gamerdinger, M., E. Deuerling, 2014, Trigger factor flexibility, Science, 344:590-591.

Ge, H., H. Qian, 2011, Non-equilibrium phase transitions in mesoscopic biochemical systems: from stochastic to nonlinear dynamics and beyond, Interface, 8:107-116.

Geronimus, A., M. Hicken, D. Keene, J. Bound, 2004, 'Weathering' and age patterns of allostatic load scores among Blacks and Whites in the United States, American Nournal of Public Health, 96:826-833.

Gilbert, P., 2001, Evolutionary approaches to psychopathology: the role of natural defenses, Australian and New Zeland Journal of Psychiatry, 35:17-27.

Giulivi, C., Y. Zhang, A. Omanska-Klusek, C. Ross-inta, S. Wong, I. Hertz-Picciotto, F. Tassone, I. Pessah, 2010, Mitochondrial dysfunction in autism, JAMA, 304:2389-2395.

Glass, T., M. McAtee, 2006, Behavioral science at the crossroads of public health: extending horizons, envisioning the future, Social Science and Medicine, 62:1650-1671.

Glazebrook, J.F., R. Wallace, 2009, Rate distortion manifolds as model spaces for cognitive information, Informatica, 33:309-346.

Goh, S., Z. Dong, Y. Zhang, S. DiMauro, B. Peterson, 2014, Mitochondrial dysfunction as a neurobiological subtype of Autism Spectrum Disorder, JAMA Psychiatry doi:10.1001/jamapsychiatry.2014.179.

Goldschmidt, L. *et al.* 2010, Identifying the amylome, proteins capable of forming amyloid-like fibrils, PNAS, 107:3487-3492.

Gould, S., 2002, The Structure of Evolutionary Theory, Harvard University Press, Cambridge, MA.

Greenwald, J., R. Riek, 2010, Biology of amyloid: structure, function, and regulation, Structure, 18:1244-1260/

Haataja, L., T. Gurlo, C. Huang, P. Butler, 2008, Islet amyloid in type 2 diabetes, and the toxic oligomer hypothesis, Endocrine Reviews, 29:303-316.

Harman, D., 1956, Aging: a theory based on free radical and radiation chemistry, Journal of Gerontology, 11:298-300.

Hazewinkel, M., 2002, Index Formulas, Encyclopedia of Mathematics, Springer, New York.

Hecht, M., *et al.* 2004, De novo proteins from designed combinatorial libraries, Protein Science, 13:1711-1723.

Heine, S., 2001, Self as cultural product: an examination of East Asian and North American selves, Journal of Personality, 69:881-906.

Herrup, K., M. Carrillo, D. Schenk, et al., 2013, Beyond amyloid: getting real about nonamyloid targets in Alzhiemer's disease, Alzheimer's and Dementia, 9:452-458.

Hollan, J., J. Hutchins, D. Kirsch, 2000, Distributed cognition: toward a new foundation for human-computer interaction, ACM Transactions on Computer-Human Interaction, 7:174-196.

Holling, C., 1973, Resilience and stability of ecological systems, Annual Reviews of Ecological Systematics, 4:1-23.

Holling, C., 1992, Cross-scale morphology, geometry, and dynamics of ecosystems, Ecological Monographs, 62:447-502.

Horsthemeke, W., R. Lefever, 2006, Noise-induced Transitions, Vol. 15, Theory and Applications in Physics, Chemistry, and Biology, Springer, New York.

Houghton, C., 1975, Wreath products of groupoids, Journal of the London Mathematical Society, 10:179-188.

Jablonka, E., M. Lamb, 1995, Epigenetic Inheritance and Evolution: The Lamarckian Dimension, Oxford University Press, Oxford.

Jablonka, E., M. Lamb, 1998, Epigenetic inheritance in evolution, Journal of Evolutionary Biology, 11:159-183.

Jablonka, E., 2004, Epigenetic epidemiology, International Journal of Epidemiology, 33:929-935.

Jackson, S., F. Perkins, E. Khandor, L. Cordwell, S. Hamann, S. Buasai, 2007, Integrated health promotion strategies: a contribution to tackling current and future health challenges, Health Promotion International, 21(S1):75-83.

Jaenish, R., A. Bird, 2003, Epigenetic regulation of gene expression: how the genome integrates intrinsic and environmental signals, Nature:Genetics Supplement, 33:245-254.

Johnson-Laird, P., F. Mancini, A. Gangemi, 2006, A hyperemotional theory of psychological illness, Psychological Review, 113:822-841.

Johri, A., M.F. Beal, 2012, Mitochondrial dysfunction in neurodegenerative diseases, Perspectives in Pharmacology, 342:619-630.

Kahraman, A., 2009, The geometry and physiochemistry of protein binding sites and ligands and their detection in electron density maps. PhD dissertation, Cambridge University.

Kaufman, J., C. Poole, 2000, Looking back on 'Causal Thinking in the Health Sciences', Annual Reviews of Public Health, 21:101-119.

Khasminskii, R., 2012, Stochastic Stability of Differential Equations, Second Edition, Springer, New York.

Kirkwood, T., 1977, Evolution of aging, Nature, 270:301-304.

Kim, W., M. Hecht, 2006, Generic hydrophobic residues are sufficient to promote aggregation of the Alzheimer's $A\beta 42$ peptide, PNAS, 103:552-557.

Kim, Y., M. Hipp, A. Hayer-Hartl, F. Hartl, 2013, Molecular chaperone functions in protein folding and proteostasis, Annual Reviews in Biochemistry, 82:323-355.

Kitano, H., 2004, Biological robustness, Nature Genetics, 5:826-837.

Kleene, S., 1956, Representations of events in nerve sets and finite automata.

In Shannon, C., (ed.), Automata Studies, vol. 3-41, Princeton University Press, Princeton, NJ.

Kleinman, A., B. Good, 1985, Culture and Depression, Califorina University Press, Berkeley.

Kleinman, A., A. Cohen, 1997, Psychiatry's global challenge, Scientific American, 276(3):86-89.

Kolb, H., T. Mandrup-Paulsen, 2010, The global diabetes epidemic as a consequence of lifestyle-induced low-grade inflammation, Diabetologic, 53:10-20.

Kozen, D., 1997, Automata and Computability, Springer, New York.

Laganowsky, A., E. Reading, et al., 2014, Membrane proteins bind lipids selectively to modulate their structure and function, Nature, 510:172-175.

Landau, L., E. Lifshitz, 2007, Statistical Physics, Part I, Elsevier, New York.

Lawson, T., 2006, The nature of heterodox economics, Cambridge Journal of Economics, 30:483-505.

Lee, H., Y. Wei, 2012, Mitochondria and Aging. Chapter 14 in Scantena et al. (eds.), Advances in Mitochondrial Medicine, Springer, New York.

Lee, J., 2000, Introduction to Topological Manifolds, Graduate Texts in Mathematics Series, Springer, New York.

Lei, J., S. Browning, S. Mahal, A. Oelschlegel, C. Weissman, 2010, Darwinian evolution of prions in cell culture, Science, 327:869-872.

Leonteif, W., 1982, Letter, Science, 217:104-107.

Levin, S., 1989, Ecology in theory and application. In Applied Mathematical Ecology, Levin, S., T. Hallam, L. Gross (eds.), Biomathematical Texts 18, Springer, New York.

Levitt, M., C. Chothia, 1976, Structural patterns in globular proteins, Nature, 261:552-557.

Lewontin, R., The Triple Helix: Gene, Organism, and Environment, Harvard University Press, Cambridge, MA.

Lewontin, R., 2010, Not so natural selection, New York Review of Books, online.

Longuet-Higgins, H., 1963, The symmetry groups of non-rigid molecules, Molecular Physics, 6:445-460.

Lorenzini, A., T. Stamato, C. Sell, 2011, The disposable soma theory revisited, Cell Cycle, 22:3853-3856.

Markus, H., S. Kityama, 1991, Culture and the slef-implications for cognition, emotion, and motivation, Psychological Review, 98:224-253.

Martinerie, J., C. Adam, M. Le Van Quyen, 1998, Epileptic seizures can be anticipated by non-linear analysis, Nature:Medicine 4:1173-1176.

Masuda, T., R. Nisbett, 2006, Culture and change blindness, Cognitive Science, 30:381-399.

Matsumoto, Y., 2002, An Introduction to Morse Theory, Translations of Mathematical Monographs, Vol. 208, American Mathematical Society, Providence.

Maturana, H., 1970, Biology of cognition, Biological Computer Laboratory, Research Report BCL 9.0, University of Illinois, Urbana.

Maturana, H., F. Varela, 1980, Autopoiesis and Cognition, Reidel, Netherlands).

Maturana, H., F. Varela, 1992, The Tree of Knowledge, Shambhala Publications, Boston.

Maury, C.,2009, Self-propagating β-sheet polypeptide structures as prebiotic informational molecular entities: the amyloid world, Origin of Life and the Evolution of Biospheres, 39:141-150.

Mayeux, R., Y. Stern, S. Spanton, 1985, Heterogeniety in dementia of Alzheimer type: evidence of subgroups, Neurology, 35:453-461.

McCauley, L., 1993, Chaos, Dynamics, and Fractals: An Algorithmic Approach to Deterministic Chaos, Cambridge University Press, New York.

McEwen, B., T. Seeman, 1999, Protective and damaging effects of mediators of stress: Elaborating and testing the concepts of allostasis and allostatic load, Annuals of the New York Academy of Sciences, 896:30-47.

Mezei, J., J. Wright J., 1967, Algebraic automata and context-free sets, Information and Control, 11:3-29.

Miller, A., V. Maletic, C. Raison, 2009, Inflammation and its discontents: the role of cytokines in the pathophysiology of major depression, Biological Psychiatry, 65:732-774.

Mindwood, K., L. Valenick Williams, J. Schwarzbauer, 2004, Tissue repair and the dynamics of the extracellular matrix, International Journal of Biochemistry and Cell Biology, 36:1031-1037.

Nair, G., F. Fagnani, S. Zampieri, R. Evans, 2007, Feedback control under data rate constraints: an overview. Proceedings of the IEEE, 95:108-137.

Nisbett, R., K. Peng, C. Incheol, A. Norenzayan, 2001, Culture and systems of thought: holistic vs. analytic cognition, Psychological Review, 108:291-310.

Nisbett, R., Y. Miyamoto, 2005, The influence of culture: holistic versus analytic perception, TRENDS in Cognitive Science, 9:467-473.

Nunney, L., 1999, Lineage selection and the evolution of multistage carcinogenesis, Proceedings of the Royal Society B., 266:493-498.

Odling-Smee, F., K. Laland, M. Feldman, 2003, Niche Construction: The Neglected Process in Evolution, Princeton University Press, Princeton, NJ.

O'Nullain, S., 2008, Code and context in gene expression, cognition, and consciousness. In Barbiere, M., (ed.), The Codes of Life: The Rules of Macroevolution, Springer, New York, pp.347-356.

Overduin, M., A. Furnham, 2012, Assessing obsessive-compulsive disorder (OCD): a review of self-report measures, Journal of Obsessive-Compulsive and Related Disorders, 1:312-324.

Oyelaran, O., Q. Li, D. Farnsworth, J. Gildersleeve, 2009, Microarrays with varying carbohydrate density reveal distinct subpopulations of serum antibodies, Journal of Proteome Research, 8:3529-3538.

Palmieri, L., A. Persico, 2010, Mitochondrial dysfunction in autism spectrum disorders: Cause or effect?, Biochimica et Biophysica Acta, 1797:1130-1137.

Panskepp, J., 2003, At the interface of the affective, behavioral, and cognitive neurosciences: decoding the emotional feelings of the brain, Brain and Cognition, 52:4-14.

Park, C., N. Larsson, 2011, Mitochondrial DNA mutations in disease and aging, Journal of Cell Biology, 193:809-818.

Pettini, M., 2007, Geometry and Topology in Hamiltonian Dynamics, Springer, New York.

Pielou, E., 1977, Mathematical Ecology, Wiley, New York.

Prabakaran, S., J. Swatton, M. Ryan, S. Huffaker, J. Huang, J. Griffin, M. Wayland et al., 2004, Mitochondrial dysfunction in schizophrenia: evidence for compromised brain metabolism and oxidative stress, Molecular Psychiatry 9:6840697.

Pretzel, O., 1996, Error-Correcting Codes and Finite Fields, Clarendon Press, Oxford.

Protter, P., 1990, Stochastic Integration and Differential Equations, Springer, New York.

Pujol, A., R. Mosca, J. Farres, P. Aloy, 2009, Unveiling the role of network and systems biology in drug discovery, Trends in Pharmacological Sciences, 31:115-123.

Quastler, H. (ed.), 1954, Information Theory in Biology, University of Illinois Press, Urbana.

Querfurth, H., F. LaFerla, 2010, Alzheimer's disease, New England Journal of Medicine, 362:329-344.

Qiu, C., M. Kivipelto, E. von Strauss, 2009, Epidemiology of Alzheimer's disease: occurrence ,determinants ,and strategies toward intervention, Dialogues in Clinical Neuroscience, 11:111-128.

Rau, H., T. Elbert, 2001, Psychophysiology of arterial baroreceptors and the etiology of hypertension, Biological Psychology, 57:179-201.

Richerson, P., R. Boyd, 2006, Not By Genes Alone: How Culture Transformed Human Evolution, Chicago University Press, Chicago, IL.

Richie, K., J. Touchon, 1992, Heterogeneity in senile dementia of the Alzheimer type: individual differences, progressive deterioration, or clinical subtypes? Journal of Clinical Epidemiology, 45:1391-1398.

Riesner, D., 2003, Biochemistry and structure of PrP^C and PrP^{Sc}, British Medical Bulletin, 66:21-33.

Ringel, G., J. Young, J., 1968, Solutions of the Heawood map-coloring problem, PNAS, 60:438-445.

Rockafellar, R., 1970, Convex Analysis, Princeton University Press, Princeton.

Roman, S., 1997, Introduction to Coding and Information Theory, Springer, New York.

Rossignol, R., R. Frye, 2010, Mitochondrial dysfunction in autism spectrum disorders: a systematic review and meta-analysis, Molecular Psyhciatry, 17:290-314.

Rossignol, D., R. Frye, 2014, Evidence linking oxidative stress, mitochondrial dysfunction, and inflammation in the brain of individuals with autism, Frontiers in Physiology, 5:150.

Sabate, R., M. Gallardo, J. Estelrich, 2003, An autocatalytic reaction as a model for the kinetics of the aggregation of β-amyloid, Biopolymers (Peptide Science), 71:190-195.

Saio, T., Guan, X., Rossi, P., et al., Structural basis for protein antiaggregation activity of the Trigger Factor chaperone, Science, 344:597 (2014).

Sawaya, M., et al., 2007, Atomic structures of amyloid cross-β splines reveal varied steric zippers, Nature, 447:453-457.

Scaglia, F., 2010, The role of mitochondrial dysfunction in psychiatric disease, Developmental Disabilities Research Reviews, 16:136-143.

Scherrer, K., J. Jost, 2007a, The gene and the genon concept: a functional and information-theoretic analysis, Molecular Systems Biology, 3:87-95.

Scherrer, K., J. Jost, 2007b, Gene and genon concept: coding versus regulation, Theory in Bioscience, 126:65-113.

Scheuner, D., R. Kaufman, 2008, The unfolded protein response: a pathway that links insulin demand with β-cell failure and diabetes, Endocrine Reviews, 29:317-333.

Schneider, T., 2010, A brief review of molecular information theory, Nano Communication Networks, 1:173-180.

Schwabe, L., K. Obermayer, 2002, Rapid adaptation and efficient coding, BioSystems, 67:239-244.

Sergeant, C., S. Dehaene, 2004, Is consciousness a gradual phenomenon? Evidence for an all-or-none bifurcation during the attentional blink, Psychological Science, 15:720-725.

Shao, L., M. Martin, S. Watson, A. Schatzberg, H. Akil, R. Myers, E. Jones, W. Bunney, M. Vawter, 2008, Mitochondrial involvement in psychiatric disorders, Trends in Molecular Medicine 40:281-295.

Shannon, C., 1959, Coding theorems for a discrete source with a fidelity criterion, Institute of Radio Engineers International Convention Record Vol. 7, 142-163.

Shirkov, D., V. Kovalev, 2001, The Bogoliubov renormalization group and solution symmetry in mathematical physics, Physics Reports, 352:219-249.

Singh-Manoux, A., N. Adler, M. Marmot, 2003, Subjective social status: its determinants and its association with measures of ill-health in the Whitehall II study, Social Science and Medicine, 56:1321-1333.

Smith, E., 2011, et al., Nonequilibrium phase transitions in biomolecular signal transduction, Physical Review E, 84:051917.

Stohr, J., et al., 2014, Distinct synthetic $A\beta$ prion strains producing different amyloid deposits in bigenic mice, PNAS, 111:10329-10334.

Stuart A., J. Ord, 1994, Kendall's Advanced Thoery of Statistics, Sixth Edition, Hodder-Arnold, London.

Susser, M., 1973, Causal Thinking in the Health Sciences: Concepts and Strategies of Epidemiology, Oxford University Press, New York.

Swerdlow, R., et al., 2010, The Alzheimer's disease mitochondrial cascade hypothesis, Journal of Alzheimer's Disease 20:S265-S279.

Thayer, J., B. Friedman, 2002, Stop that! Inhibition, sensitization, and their neurovisceral concomitants, Scandinavian Journal of Psychology, 43:123-130.

Timberlake, W., 1994, Behavior systems, associationism, and Pavlovian conditioning, Psychonomic Bulletin, Rev. 1, 405-420.

Tishby, N., D. Polani, 2011, Information theory of decisions and actions, Chapter 19 in Cutsuridis et al. (eds.), Perception-Action Cycle: Models, Architectures, and Hardware, Springer Series in Cognitive and Neural Systems 1, New York.

Tlusty, T., 2007, A model for the emergence of the genetic code as a transition in a noisy information channel, Journal of Theoretical Biology, 249:331-342.

Tlusty, T., 2008, A simple model for the evolution of molecular codes driven by the interplay of accuracy, diversity and cost, Phys. Biol., 5:016001; Casting polymer nets to optimize noisy molecular codes, PNAS, 105:8238-8243.

Tomkins, G., 1975, The metabolic code, Science, 189:760-763.

Tompa, P., M. Fuxreiter, 2008, Fuzzy complexes: polymorphism and structural disorder in protein-protein interactions, Trends in Biochemical Science, 33:1-8.

Turner, B., 2000, Histone acetylation and an epigenetic code, Bioessays, 22:836-845.

van Lint, J., 1999, Introduction to coding Theory, Springer, New York.

Van den Broeck, C., J. Parrondo, R. Toral, 1994, Noise-induced nonequilibrium phase transition, Physical Review Letters, 73:3395-3398.

Van den Broeck, C., J. Parrondo, J., R. Toral, R. Kawai, 1997, Nonequilibrium phase transitions induced by multiplicative noise, Physical Review E, 55:4084-4094.

Varela, F., E. Thompson, E. Rosch, 1991, The Embodied Mind: Cognitive Science and Human Experience, MIT Press.

Verduzco-Flores, S., Ermentrout, B., Bodner, M., 2012, Modeling neuropathologies as disruption of normal sequence generation in working memory models, *Neural Networks*. 27, 21-31.

Wallace D., R. Wallace, 2000, Life and death in Upper Manhattan and the Bronx: toward evolutionary perspectives on catastrophic social change, Environment and Planning A, 32:1245-1266.

Wallace, D.C., 2005, A mitocondrial paradigm of metabolic and degenerative diseases, aging, and cancer: a dawn for evolutionary medicine, Annual Reviews of Genetics, 39:359-407.

Wallace, D.C., 2010, Mitochondrial DNA mutations and aging, Environmental Molecular Mutagens, 51:440-450.

Wallace, R., 2000, Language and coherent neural amplification in hierarchical systems, International Journal of Bifurcation and Chaos, 10:493-502.

Wallace, R., 2005a, Consciousness: A Mathematical Treatment of the Global Neuronal Workspace Model, Springer, New York.

Wallace, R., 2005b, A global workspace perspective on mental disorders, Theoretical Biology and Medical Modelling, 2:49.

Wallace, R., 2007, Culture and inattentional blindness: a global workspace perspective, Journal of Theoretical Biology, 245:378-390.

Wallace, R., 2010a, A scientific open season, Physics of Life Reviews, 7:377-378.

Wallace, R., 2010b, Expanding the modern synthesis, Comptes Rendus Biologies, 333:701-709.

Wallace, R., 2011, Multifunction moonlighting and intrinsically disordered proteins: information catalysis, non-rigid molecule symmetries and the 'logic gate' spectrum, Comptes Rendus Chimie, 14:1117-1121.

Wallace, R., 2012a, Consciousness, crosstalk, and the mereological fallacy: an evolutionary perspective, Physics of Life Reviews, 9:426-453.

Wallace, R., 2012b, Spontaneous symmetry breaking in a non-rigid molecule approach to intrinsically disordered proteins, Molecular BioSystems 8:374-377.

Wallace, R., 2012c, Extending Tlusty's rate distortion index theorem method to the glycome: do even 'low level' biochemical phenomena require sophisticated cognitive paradigms? BioSystems, 107:145-152.

Wallace, R., 2013, A new formal approach to evolutionary processes in socioeconomic systems, Journal of Evolutionary Economics, 23:1-15.

Wallace, R., 2014a, Cognition and biology: perspectives from information theory, Cognitive Processing 15:1-12.

Wallace, R., 2014b, Statistical models of critical phenomena in fuzzy biocognition, BioSystems, 117:54-59.

Wallace, R., 2014c, A new formal perspective on 'Cambrian Explosions', Comptes Rendus Biologie, 337:1-5.

Wallace, R., 2014d, Dynamic statistical models of biological cognition: insights from communications theory. In press, Connection Science.

Wallace, R., 2015, An Ecosystem Approach to Economic Stabilization: Escaping the neoliberal wilderness, Routletge Series on Heterodox Economics, London.

Wallace, R., R.G. Wallace, 2002, Immune cognition and vaccine strategy: beyond genomics, Microbes and Infection, 4:521-527.

Wallace, R., D. Wallace, R.G. Wallace, 2003, Toward cultural oncology: the evolutionary information dynamics of cancer, Open Systems and Information Dynamics, 10:159-181.

Wallace, R., D. Wallace, 2004, Structured psychosocial stress and therapeutic failure, Journal of Biological Systems 12:335-369.

Wallace, R., M. Fullilove, 2008, Collective Consciousness and its Discontents, Springer, New York.

Wallace, R., D. Wallace, 2008, Punctuated equilibrium in statistical models of generalized coevolutionary resilience: how sudden ecosystem transitions can entrain both phenotype expression and Darwinian selection, Transactions on Computational Systems Biology IX, LNBI 5121:23-85.

Wallace, R., D. Wallace, 2009, Code, context, and epigenetic catalysis in gene expression, Transactions on Computational Systems Biology XI, LNBI 5750:283-334

Wallace, R., D. Wallace, 2010, Gene Expression and its Discontents: The Social Production of Chronic Disease, Springer, New York.

Wallace, R., D. Wallace, 2011, Protein Folding Disorders: From basic biology to public policy, Amazon Create Space, ISBN 9781467915946.

Wallace, R., D. Wallace, 2013, A Mathematical Approach to Multilevel, Multiscale Health Interventions: Pharmaceutical industry decline and policy response, Imperial Colleg Press, London.

Wallach, J., M. Rey, 2009, A socioeconomic analysis of obesity and diabetes in New York City, Public Health Research, Practice, and Policy, Centers for Disease Control and Prevention.

Wang, H., M. Wahlberg, A. Karp, B. Winblad, L. Fratiglioni, 2012, Psychosocial stress at work is associated with increased dementia risk in late life, Alzheimer's and Dementia, 8:114-120.

Watts, J., et al., 2014, Serial propagation of distinct Aβ prions from Alzheimer's disease patients, PNAS, 111:10323-10328.

Weinstein, A., 1996, Groupoids: unifying internal and external symmetry, Notices of the American Mathematical Association, 43:744-752.

Westermark, P., 2005, Aspects on human amyloid forms and their fibril properties, FEBS Jurnal, 272:5942-5949.

Wilson, A., S. Golonka, 2013, Embodied cognition is not what you think it is, Frontiers in Psychology, doi: 10.3389/fpsyg.2013.00058.

Wilson, K., 1971, Renormalization group and critical phenomena I. Renormalization group and the Kadanoff scaling picture, Physical Review B, 4:3174-3183.

Wilson, M., 2002, Six views of embodied cognition, Psychonomic Bulletin and Review, 9:625-636.

Wolpert, D., W. Macready, 1995, No free lunch theorems for search, Santa Fe Institute, SFI-TR-02-010.

Wolpert, D., W. Macready, 1997, No free lunch theorems for optimization IEEE Transactions on Evolutionary Computing, 1:67-82.

Yao, J. R. Irwin, L. Zhao, J. Nilsen, R. Hamilton, R. Diaz Brinton, 2009, Mitochondrial bioenergetic deficit precedes Alzheimer's pathology in female mouse model of Alzheimer's disease, Proceedings of the National Academy of Sciences, 106:14670-14675.

Yeung, R., 2008, Information Theory and Network Coding, Springer, New York.

Zhang, Q., Y. Wang, E. Huang, 2009, Changes in racial/ethnic disparities in the prevalence of type 2 diabetes by obesity level among US adults, Ethnicity and Health, 14:439-457.

Zhao, S., R. Iyengar, 2012, Systems pharmacology: network analysis to identify multiscale mechanisms of drug action, Annual Review of Pharmacology and Toxicology, 52:505-521.

Zhu, R., A. Rebirio, D. Salahub, S. Kaufmann, 2007, Studying genetic regulatory networks at the molecular level: delayed reaction stochastic models, Journal of Theoretical Biology, 246:725-745.

Zurek, W., 1985, Cosmological experiments in superfluid helium? Nature:317:505-508.

Zurek, W., 1996, The shards of broken symmetry, Nature, 382:296-298.

Index

Printed in the United States
By Bookmasters